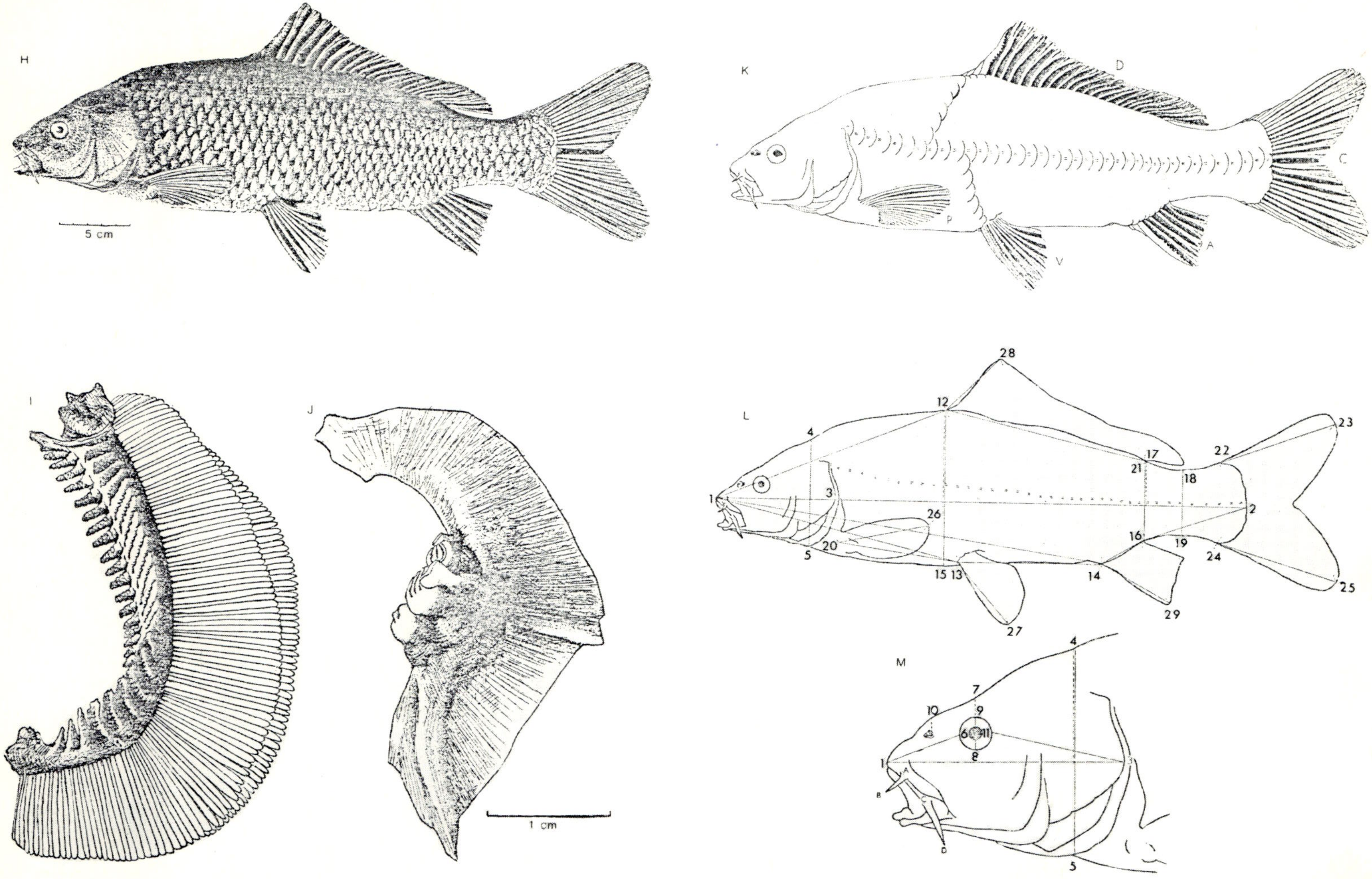
H
5 cm
K
D
C
P
V
A
I
J
1 cm
L
M

Fig. 1 Coloration of Danube wild carp and reared offspring. Top: Danube wild carp, female, 49 cm standard length, from Lesser Danube above Kolarovo, Slovakia. Centre: first generation of wild carp reared in ponds, eight-month-old fish, 15 cm long. Bottom: seven-year-old, "dwarfed" Danube wild carp, 15 cm long (photo E. K. Balon).

ROYAL ONTARIO MUSEUM

LIFE SCIENCES

MISCELLANEOUS PUBLICATION

EUGENE K. BALON

Domestication of the Carp *Cyprinus carpio* L.

Suggested citation: Roy.Ont.Mus. Life Sci. Misc. Pub.
ISBN: 0-88854-147-3
Publication date: 29 April 1974

ROYAL ONTARIO MUSEUM

PUBLICATIONS IN LIFE SCIENCES

The Royal Ontario Museum publishes three series in the Life Sciences:

LIFE SCIENCES CONTRIBUTIONS, a numbered series of original scientific publications, including monographic works.

LIFE SCIENCES OCCASIONAL PAPERS, a numbered series of original scientific publications, primarily short and usually of taxonomic significance.

LIFE SCIENCES MISCELLANEOUS PUBLICATIONS, an unnumbered series of publications of varied subject matter and format.

All manuscripts considered for publication are subject to the scrutiny and editorial policies of the Life Sciences Editorial Board, and to review by persons outside the Museum staff who are authorities in the particular field involved.

E. K. BALON is Associate Professor, Department of Zoology, University of Guelph, and Research Associate, Department of Ichthyology and Herpetology, Royal Ontario Museum.

PRICE: $3.00

100 Queen's Park, Toronto, Canada

PRINTED AT THE BRYANT PRESS LIMITED

Colin Bertram (Oryx, 1963, p.214)	"*Preservation by domestication in an alien environment is far better than extinction in the natural habitat.*"
I disagree	*In spite of commonly held views that domestication preserves—it does not—it alters.*
H. Epstein (R. L. Smith's "The Ecology of Man: . . .", 1972, p.91)	"*Domestication changed the life of the beast, the character of the animal, and its anatomy and physiology.*"

The cover: Fishermen of the wild carp on the Danube River near Komarno one hundred years ago (reprinted from an etching in O. Herman, A Magyar halaszat könyve, Budapest, 1887).

Contents

Domestication of the Carp *Cyprinus carpio* L.

Abstract

Historical, zoogeographical, morphological, and physiological information were utilized in explaining the origins and history of domestication of the carp. The wild carp ancestor apparently originated in central Asia and spread naturally east into China and west as far as the Danube River. Evidence suggests that the Romans first cultured carp collected from the Danube, and the tradition of the *piscinae* was continued in monasteries throughout the Middle Ages. A starvation experiment demonstrated the hardiness of Danube wild carp and confirmed its suitability for domestication. Domestication of the carp in China was first independent of similar activities in Europe. The strong evidence pointing to rheophilic Danube wild carp as ancestral to modern domestic varieties suggests the importance of preserving this natural population for further use.

Résumé

Pour expliquer les origines et l'historique de la domestication de la carpe on a mis en oeuvre différentes sources de renseignements; l'histoire, la zoogéographie, la morphologie et la physiologie.

On trouve l'ancêtre de la carpe (à l'état sauvage) apparemment d'abord en Asie centrale, puis graduellement vers l'est jusqu'en Chine et vers l'ouest jusqu'au Danube. Il y a tout lieu de croire que c'étaient les Romains qui firent les premiers l'élevage de la carpe qu'ils prenaient du Danube, et on continua la tradition des *piscinae* dans les monastères pendant le Moyen Age. On a tenté de démontrer par des expériences avec la carpe sauvage du Danube, la résistance de celle-ci à un régime de jeûne forcé, et ainsi on a même prouvé qu'on pourrait en faire parfaitement un élevage systématique. L'élevage de la carpe en Chine s'est développé d'abord indépendemment de celui de l'Europe. Les recherches faites nous ont fourni la quasi-certitude que la carpe sauvage réophile du Danube est bien l'ancêtre des différentes espèces modernes d'élevage, et ainsi elles mettent même au tout premier plan l'intérêt à préserver cette population naturelle pour des usages ultérieurs.

Introduction

The wild carp (*Cyprinus carpio carpio*) is the predecessor of the domesticated or pond carp about whose vicissitudes we have less information than about the extinct auroch, the predecessor of cattle. An attempt is made, therefore, to document its evolutionary and domestic history.

During the past 400 years the domestic carp has been intensively cultured in Europe and introduced into many countries around the world. The origin, however, of this valuable commercial species has never been satisfactorily explained. The writing of this paper was stimulated by recent publications on the history of the carp in Europe (by Rudziński, 1962, among others), Australia (Butcher, 1967), and North America (among others, McCrimmon, 1968) that repeated old, unfounded ideas regarding the origin of this fish and also omitted details of its prehistory and early history (Vooren, 1972).

Results of a number of studies (Balon and Mišik, 1956; Mišik, 1958; Mišik and Tuča, 1965; Balon, 1957, 1958a, 1958b; Bastl, 1961; Rudziński, 1961) bearing on wild carp domestication are summarized here in an attempt to relate biological results to facts gleaned from zoogeography, history, and archaeology. As these data are meagre and their interpretation often ambiguous or subjective, the origin of domestic carp may not be solved definitively even in this essay. If my hypothesis, however, encourages more research and assists archaeology and social anthropology to interpret their newly discovered artifacts and data, the goal of this paper will have been accomplished.

An earlier version of this study appeared in the mimeographed "Works of the Laboratory of Fishery Research in Bratislava" (Balon, 1969). Presented here is a revised and amended version, with a new section on an experiment in starvation. The experiment was a result of an incidental activity, and though lacking details of a strictly scientific study, is worthy of record.

Materials and Methods

The first numerous wild carp in the piedmont zone of the Danube River were collected from a spawning school in May and June of 1955. Although I sampled individual specimens a year earlier (Balon and Mišik, 1956), the intensified search resulted in finding the spawning school near the village of Kolarovo. These fish formed the main material for the age and growth study (Balon, 1957) and for the taxonomic revision (Mišik, 1958). Although initially added to the collection of the Laboratory of Fishery Research (Slovak Agricultural Academy), most of them were later transferred to and made a part of the collection of the Slovak National Museum in Bratislava. Some specimens from successive spawning and developmental experiments (Balon, 1958a), pond culture (Bastl, 1961) and from the experiment in starvation were also preserved and stored in both collections. These collections can now be considered major retainers of preserved rheophilic wild carp of the Danube River.

The carp were kept in aquaria for an experiment in starvation (date of hatching—31 May 1956). Individual specimens were successively selected for preservation or study of scales: first and second juveniles were preserved on 13 October 1956 (standard length [SL] 44 and 61 mm, weight [w] 2 and 5 g respectively); third specimen, a mature female, was preserved 19 March 1959 (SL 128 mm, w 56 g); from the fourth specimen scales were studied on 20 June 1958 (SL 129 mm, w 53 g) and the fish was preserved as a mature male on 19 March 1959 (SL 144 mm, w 67 g); from the fifth specimen scales were studied on 20 June 1958 (when SL 120 mm, w 48 g) and the fish was preserved as a mature male on 9 December 1960 (when SL 134 mm—the fish was unsuitable for most measurements because it jumped out of the aquarium and was found dehydrated); from the sixth and seventh specimen scales were studied on 13 October 1961 (SL 150 and 148 mm, w 69 and 56 g respectively) and on 20 June 1958 (when SL 138 and 146 mm, w 63 and 80 g respectively), the sixth fish was preserved as a mature female on 26 April 1962 (SL 148 mm, w 47 g), the seventh fish as a mature female on 6 January 1963 (SL 148 mm, w 68 g).

The relative growth indices used in Table 1 were explained in an earlier study (Balon, 1964, 1972). The identification of sex and usage of meristic and morphometric characters (Table 2) is defined in detail in Mišik's (1958) study; the indices in Table 3 in Rudziński's (1961) and Steffens' (1964) studies.

Figures on the inside covers give some illustration of the extensive initial studies of the wild carp. They are reprinted here primarily for the benefit of the English language reader, and they explain the counts, measurements and usage of terms.

INSIDE THE FRONT AND BACK COVERS

Front cover: A–G some developmental stages of the wild carp (from Balon, 1958a); A—embryo at time of hatching and its main respiratory organs—ducti Cuvieri in anterior part of the yolk sac; C—feeding larva 12 days old and 11 mm long with well developed (D) external lateral line sensors (cupulae); E—full grown larva 20 days old and18 mm long; F—the appearance of first scales on a 15 mm long larva, and G—a 22 mm long juvenile. H–M some characters of an adult wild carp (from Mišik, 1958); H—the type wild carp from the Danube River near Medveďovo (21.5.1954, Slovak National Museum #171), I—its first gill arch and J—pharyngeal bone; K—sketch indicating meristic characters; L, M—sketch indicating morphometric characters (numbers refer to characters listed in Table 2).

Back cover: A—scaled morph, B—line or heavily scaled mirror morph, C—lesser scaled mirror morph, and D—scaleless or leather morph of the domestic carp (after Wunder, 1949); E—the type wild carp (from Balon, 1967b); F—the rate of growth of wild carp (from Balon, 1957); G—the wild and H—the domestic carp (after Holčik and Hensel, 1972).

Paleogeographical Distribution of Carp

Studies proving the occurrence of the carp in Western Europe as early as the Tertiary are based primarily on scales found in preglacial freshwater strata in northern Germany (Nehring, 1883, after Zaunick, 1925) and at pile dwelling sites in Switzerland. Ruetimeyer's (1860) claim that he found scales of carp in remnants of pile dwellings was later contradicted by Forel (1904). Forel based his opinion on the information of Studer (in Zaunick, 1925) who identified the scales as those of *Abramis brama*. Thus, not only is the identity of preglacial remnants doubtful, but their age is unconfirmed. There is no further true paleontological evidence. Consequently, the occurrence of wild carp in preglacial Europe is conjectural.

The most correct opinions seem to be those of authors (see in Mišik, 1958) who considered Asia Minor and the area of the Caspian Sea to be the origin of the wild carp. The primitive morph or subspecies of the wild carp, *Cyprinus carpio anatolicus* Hankó, 1924, with the greatest number of pharyngeal teeth, still occurs there. All morphs of the wild carp even today are highly adaptable and almost always react to changing environmental conditions by an abrupt modification of genotype (Dolzhenko, 1953; Burmakin, 1956; and others). Therefore, all morphs of wild carp may be considered to be evolutionarily young. That the primitive morph had started to occupy streams peripheral to western Asia where it originated as early as the late Pliocene is proved by remnants in "pontic" lacustrine strata (Borzenko, 1926; Hankó, 1932; Bănărescu, 1960). However, it probably did not survive the Pleistocene in that new area. Certain negative results support my hypothesis. Had the wild carp lived in the Danube refuge in the Pleistocene it would certainly have moved north to Scandinavia and the British Isles as continental glaciers retreated before the end of the period of interfluvial connections as did, for example, *Abramis brama*, *Esox lucius*, *Rutilus rutilus*, *Scardinius erythrophthalmus*, *Tinca tinca*, *Blicca bjoerkna*, and *Cobitis taenia* (Thienemann, 1950). Carp also did not penetrate (e.g. via the Bering land bridge) to North America in the Pleistocene, at a time when this connection enabled *Esox lucius* to do so (Lindberg, 1962, Crossman and Harington, 1970). Therefore the carp likely had not reached the Far East by the end of the Pliocene. Thus, the ancestors of recent carp probably evolved in the Caspian area and spread from there to western Europe and to China as late as the last postglacial period.

Except for the Danube River, the natural occurrence of carp in waters of Europe as early as the beginning of the Christian era is not probable. Hence Ausonius (A.D. 310–393; 1933) did not mention carp in the fauna of the Rhine and Mosel rivers in the fourth century. Later records of sporadic occurrence of carp in rivers may be explained by individuals having escaped from ponds, though Długosz (1863–1887) did not mention the presence of carp in Polish waters even as late as the 15th century.

The ancestor of contemporary wild carp evidently evolved in the area of the Caspian Sea at the end of the Pleistocene. Under conditions of the postglacial thermal optimum some strains spread as far as the Black Sea area, the Aral system, into eastern Asia and appeared in the Danube River system about 8,000–10,000 years ago, that is, somewhat sooner than did the goldfish, *Carassius auratus gibelio*, which today penetrates the Danube River (Hensel, 1971). The westward expansion of the range of that species was probably similar to that of the wild carp (Balon, 1962, 1963a).

Jordan and Evermann (1896–1900, 1902) and Burns (1966) were thus mistaken to consider the carp to be an Asian species that was introduced in Europe before 1758 and was named after the Island of Cyprus, which they considered to be the centre of its distribution. Also in error was Butcher (1967) when stating, "Whatever changes in the environment were brought about following the introduction of this fish into Europe have been lost apparently in antiquity."

According to ancient stories and myths, the Great Schutt Island (in the upper part of the Middle Danube) was surrounded by a "great number of golden carp (*Cyprinus auratus*) that enabled even the poorest people to make a living; yes, there were times when the fishermen gave them away as gifts" (translated from Khin, 1930). Then why did Jordan (and Evermann, 1896–1902), Thienemann (1950), Maar (1960), Vooren (1972), and so many others believe that the carp reached Rome via Greece from China? Possibly even the *carpio* of Plinius (A.D. 24–79; 1635) was not an unknown seafish but the wild carp; then, of course, the specific name could be from the Greek root Karpos, meaning fruit.

HISTORICAL EVIDENCE OF DANUBIAN DISTRIBUTION

Natural occurrence of the wild carp in the Danube River is supported by the evidence of Leonhardt (1906), who however hypothesized that the wild carp arrived there from its northern distribution area before the formation of the continental pack-ice. Leonhardt (1906) assented that names used for the carp in local areas and the ultimate derivation of the scientific name can tell much about the historical distribution of that fish; he advanced a probable explanation of the generic and specific name of the carp. The generic name is from latinized Greek and the species name comes from the Celtic colloquial name for the fish. "Kyprinos" or "Kyprianos" was the name given by Aristotle (384–322 B.C.; 1862) to this fish and was probably derived from "Cypris," a secondary name of the Goddess of Love, Aphrodite—perhaps because the high fertility of the carp was known even then. Later the name was latinized to "Cyprinus," probably by Pliny. Belon (1555) mentioned that he encountered the carp in the land of the Etolic Greeks, who called it "Kiprinos."

The specific name, however, probably originated later and as "carpa" appeared for the first time in the works of Cassiodorus (A.D. 490–585; 1626) in the 6th century. The old Romans called it *cyprinus*, a name that spread with the fish throughout Europe. The lack of indigenous local names for this fish to the west and north of the Danube River may be considered to be further evidence that the name originated there with the appearance of the species. The Latin name of the fish is Celtic in origin and dates from the time when Celtic tribes inhabited the present eastern Austrian and Czechoslovak–Hungarian territory of the Danube River. From the ancient "charpho," "carfo" and "charofo" the name gradually changed to the present "carpo." Celts were certainly familiar with the carp at the time they settled in the Danube area, from where the name spread with them or with later introductions of the fish. Everywhere the Celts lived, the root of the name "carp" occurs—"carpe" in France; "carpio" or "carpione" in Italy; "carp" in England; "carpe" in Spain; "Karpfen" in Germany; "karp" in Poland and Russia; and "kapr" and "Kapor" in Czechoslovakia. In the region of the Danube River, where it originated, however, this fish is called "ponty" (Hungarian), "sharan" (Serbian), "saran" (Bulgarian), "crap," "saran," "ciortan," "ciuciu" or "olocari" (Rumanian), "husgun" (Turkish), "korop," "sharan," "podrojek" (Ukrainian), "sazan" (on the Volga River), and "kalynshyr" (Kirghizia).

Cassiodorus (A.D. 490–585; 1626) wrote about shipments of Danube carp to Ravenna in Italy for the table of King Theodorus, and his note may also serve as evidence of the Danubian origin of the western domestic carp. Consequently, there is no basis for later opinions which were copied without critical comment from one book to the next, that had the wild carp originate in China or eastern Asia and suggested that European pond-culture was, so to speak, the continuation of its earlier culture in China (Maar, 1960; Steffens, 1967). These concepts were supported by some authors (Dubravius, 1547; von Hohberg, 1687) in the Middle Ages who wrote at a time when pond culture was spreading but completely misplaced the origin of the fish. They found, correctly, that carp existed in European ponds before occurring in rivers but erroneously included the Danube among the rivers. According to von Hohberg (1687, p. 582) for example:

Die See-Karpfen (oder die in den Flüssen gefangen werden) hält man auch für besser (und wiewol sie gewöhnlich in den Flüssen und Strömen nich wohnen) geschiehet es doch vielmal (dass die abgebrochen Teiche ihre Fische mit samt dem Wasser dahin überlassen müssen) wie man an dem Teyafluss spühren kan (der seine aus den Teichen enpfangene Karpffen der March) die March aber bei Töben der Donau mitt-heilet (daher auch alzwischen Pressburg und Töben viel Karpffen in der Donau zu finden sind; sie gelangen zu grossem Alter) wiewol sie in vier und fünf Jahren zu Speise am dienlichsten.

The lake carp (or that caught in rivers) is considered to be better, and although it ordinarily does not inhabit rivers and streams, it often finds its way into rivers with water from ponds with broken dams. Carp, released from ponds, are transferred first to the Dyje River and then to the Morava River, which joins the Danube at Devin. In the Danube River between Bratislava and Devin occur many carp. Individuals reach a great age, although they are most edible when they are 4 to 5 years old.

That individuals of the pond morph that escaped into rivers soon changed from a short, scaleless morph into an elongate, scaled morph similar to the wild carp supports von Hohberg's (1687) statement. A change in body shape is also known (Tchen, 1956) in varieties of the goldfish (*Carassius auratus auratus*). Although observers in the Middle Ages considered the numerous carp at the mouth of the Morava River near Devin to have escaped from ponds, it was certainly the endemic, wild form of the carp that was abundant in that part of the river, which has a large flood zone needed for successful spawning of carp.

Taxonomy and Evolution

Recent native carp from the Danube River along the Czechoslovakia–Hungary border are an ancient morph and ecologically isolated from escaped or introduced pond carp (Mišik, 1958). But of all wild morphs of the carp—from the Danube River, Aral Sea, and central and eastern Asia—that in the Danube is nearest to the domestic carp (Balon, 1957, 1958a; Mišik, 1958). In other words, we consider the wild carp of the Danube River to be not a form of pond carp which became feral after it was released into the river as is generally accepted, but as a true wild morph related to the domestic pond carp in a way similar to that of the ancestral auroch to cattle. In nature the wild carp does not hybridize with the domestic carp. Domestic carp are never found in schools of wild carp, and single, domestic carp caught within the spawning grounds of wild carp were never ready to spawn. Domestic carp which gradually change in rivers into a scaled, elongate morph similar to the wild carp, were normally easily recognized by a notch and hump posterior to the head; in native wild carp the transition between the top of the head and the back (nape) is slightly if at all marked. Numerous breeding experiments (Tuča, 1958; Bastl, 1961, 1962; Rudziński, 1961, 1962; Mišik and Tuča, 1965) proved the genetic distinctness of this morph.

After a biometric study of wild Danube carp and after comparison with wild carp from other regions, Mišik (1958) demonstrated that wild carp can be divided into three groups:

1. European wild carp (*Cyprinus carpio carpio*) represented by the population of the Danube River system and designed as nominate subspecies;
2. Eastern Asian wild carp (*Cyprinus carpio haematopterus*) from Siberia and China;
3. Wild carp "from the Aral Sea and from other central Asian regions which in some ways are more closely related to the European ones, in others to east Asian carp, which at the same time are mutually substantially different" (translated from Mišik, 1958, p. 106).

Concerning the third group Mišik (loc. cit.) concluded (my translation): "Reasons for these convergent and divergent changes in morphological features of the central Asian carp—in my opinion—are to be sought in the geographic conditions of the area, which may include specific peculiarities of the geography of both Europe and eastern Asia. It is particularly those specific characters that may show the orientation of the evolutionary divergence of central Asian carp. That is why they may be considered as inter-digitating, allopatric populations, which—if they are geographically isolated from the subspecies *C. c. haematopterus* and *C. c. carpio*—can be considered as separate geographical morphs with numerous, ecological modifications." Mišik's divisions, however, may be interpreted also as follows:

The differentiation of the wild carp in the region of the Aral Sea and central Asia was influenced by an earlier geological age; if central Asia is considered to be the origin of the wild carp, descendants of that original strain had a longer time, the entire Pleistocene, to evolve in isolation as compared to carp that emigrated eastward and westward in the last postglacial period. Even though strains later integrated, more morphs may exist today in western and central Asia than elsewhere. Regarding the low degree of differentiation, which nowhere reaches the species level, the wild carp is a geologically young species—dating at the most from the Pleistocene. From the western and central Asian area of first occurrence and origin of the wild carp, part of one strain later invaded the west, whereas part of another one has penetrated to the east. Consequently, *Cyprinus carpio carpio* and *C. c. haematopterus* exist today through their range as relatively poorly differentiated taxa. The degree of relationship between the diverse strains of central Asia and domestic stocks has yet to be established.

Dwarfed Wild Carp and Their Viability

Using wild adults selected on specially devised criteria (Mišik and Tuča, 1965), morpho-ecological characters were studied first (Balon, 1958a, 1958b), and juveniles from the same spawn were used to evaluate the extent of changes following culture in southern (Bastl, 1961) and northern (Rudziński, 1961) European ponds (Slovakia–Poland). After developmental studies were terminated, some of these juveniles were maintained in aquaria in the laboratory and are discussed here. Though a more extensive study was originally planned, as a result of personal circumstances only a part of the data remained in my possession. In spite of this and the nine years elapsed since the termination of this experiment a record of the results may have some value.

Seven fish, hatched from eggs collected after carp spawning in a pond on 29 May 1956, were kept in small (30 l) aquaria. Each year, from 1956 through 1963, some of the experimental fish were preserved. From the others, scales in the centre of body sides (key scales) were extracted, standard lengths and weights were recorded, and the fish were returned to the aquaria. The last fish was preserved after 7 years of continuous aquarium life. The average temperature was 21°C (range 18–24°C), although in the winter of 1958, as a result of a heating failure, the temperature was 10°C for several days.

Throughout the experimental period the fish were fed mainly with *Tubifex* sp. at the rate of 1–2 g/week in the first 2 years and 3–4 g/week thereafter. Only on a few occasions was the *Tubifex* replaced by the same weight of scraps of beef or live Cladocera. Although the diet was minimal to sustain life, in the first 2 years it was sufficient for some growth. The explanation could be that the ration provided, through that interval, some kind of nutrient more suitable for the developmental steps in question (Balon, 1971a) than for later steps, though even then the fish did not die as predicted by Vanecov for such cases (Balon, 1960a). During the 7 years of the experiment the fish behaved normally, and gonads developed each year in the last four years, although in the last five years the fish did not grow at all. These experiments demonstrated the ability of the wild carp to survive on small rations of substitute food, the ability to survive starvation. (The term "starvation" is justified by the minimal maintenance diet given and by the remarkable changes in body proportions of experimental fish.) As a consequence of starvation, however, the ossification of some skeletal elements of starved carp sets in earlier than in naturally developing carp (Balon, 1960b). The fish became asymmetrically "dwarfed" due to artificially induced achondroplasia-like condition.

Starvation affected growth of experimental fish, which never exceeded the average length and weight of the 1-year-old fish from the natural habitat (Table 1). Nonetheless, both wild and experimental animals became sexually mature in the third growing season.

Growth of experimental fish was more than six-fold less than that of wild fish in the Danube River. Furthermore, growth of experimental carp stopped in the fourth year. Differences in growth are reflected in relative indices (Table 1). The "index of average size" of the wild carp in the Danube was 7 and only 2 in experimental carp; the "index of weight growth" among wild carp from a spawning school in the Danube River was 1052, whereas for first generation reared in ponds it was 962, and among experimental fish, 1964.

Sexually-mature gonads of experimental carp were first found in a male and female 34 months old preserved in March 1959. Ovaries of the female, which measured 128 mm (standard length), weighed 0.5 g (representing 0.9% of the specimen's wet weight). The number of eggs was 3,950.

Ovaries of a female that was preserved 3 years later (in April 1962; age 5+) weighed 2.8 g (left 1.56 g; right 1.16 g) and represented 5.8% of the wet fish weight. The number of eggs, estimated by the same method (gravimetric), was 83,540. Ovaries of the last fish, a female that was 6+ years old and which was preserved in January 1963, were thin, compact strips adhering to the lateral and ventral sides of the intestine and contained large, single eggs that were irregularly distributed. New gonadal tissue had begun to develop around eggs from the previous year that had not yet been fully resorbed, a phenomenon that is not unexpected for a fish unable to spawn in an aquarium and forced to resorb the eggs for the last four years. Hence, in these experimental fish, age at first attainment of sexual maturity and number of eggs produced are not related to size but to age. Did stress conditions of starvation produce changes leading to a new type of stock dynamic (Balon, 1963b, p. 535) within the same generation?

Starving did bring forth distinct changes in body proportions (Fig. 2) and in some meristic characters. Counts of fin rays of experimental fish are within the range of those of wild carp (Mišik, 1958) inhabiting the Danube and those of the pond carp (Steffens, 1964). Scale counts and number of gillrakers are also within the range of normal variability. Pharyngeal tooth count and arrangement, however, are unusual (Table 2). Neither wild carp

TABLE 1. Absolute and relative growth values of wild carp (*Cyprinus carpio carpio*) from natural habitats, from ponds, and from aquaria*

Origin	*Values for absolute and relative growth*		*Growth seasons* 0	1	2	3	4	5	6	7	8
Wild carp of the Danube River (from Balon, 1957)	Standard length (mm)	l		147	262	331	382	408	436	462	484
	Wet weight (g)	w		56	322	742	1155	1472	1815	2150	2495
	Absolute increments (mm)	h		147	115	69	51	26	28	26	22
	Index of average size	ϕH					7.27				
	Weight rate of growth	Cw		5600j	475j	130j	56a	28a	23a	18s	16s
	Index of weight growth intensity	ϕCw				1052					
Danube wild carp in ponds, first generation (Bastl, 1962)	Standard length (mm)	l		86	237	348	393				
	Wet weight (g)	w		19	343	1078	1408				
	Absolute increments (mm)	h		86	153	109	45				
	Index of average size	ϕH			9.82						
	Weight rate of growth	Cw		1900j	1705j	214j	31a				
	Index of weight growth intensity	ϕCw			962						
Starved, dwarf wild carp in aquaria	Standard length (mm)	l		53	112	134	144	146	146	146	
	Wet weight (g)	w		4	30	53	62	64	64	64	
	Absolute increments (mm)	h		53	59	22	10	2	0	0	
	Index of average size	ϕH				2.09					
	Weight rate of growth	Cw		400j	650j	77j	17a	3a	0a	0a	
	Index of weight growth intensity	ϕCw				164					

* $h = l_i - l_{i-1}$, $\phi H = \sum_{i=1}^{i_{j+a}} h_i$, $Cw = \frac{w_i - w_{i-1}}{w_{i-1}} \cdot 100$, $\phi Cw = \sum_{i_w=1}^{i_{j+a}} Cw_i$ — where *l* is the fish standard length during *i*th year of growth, *j* for juvenile, *a* for adult, and *s* for senective period (Balon, 1971a). Weight indices are obtained by replacing the symbol *l* by *w*.

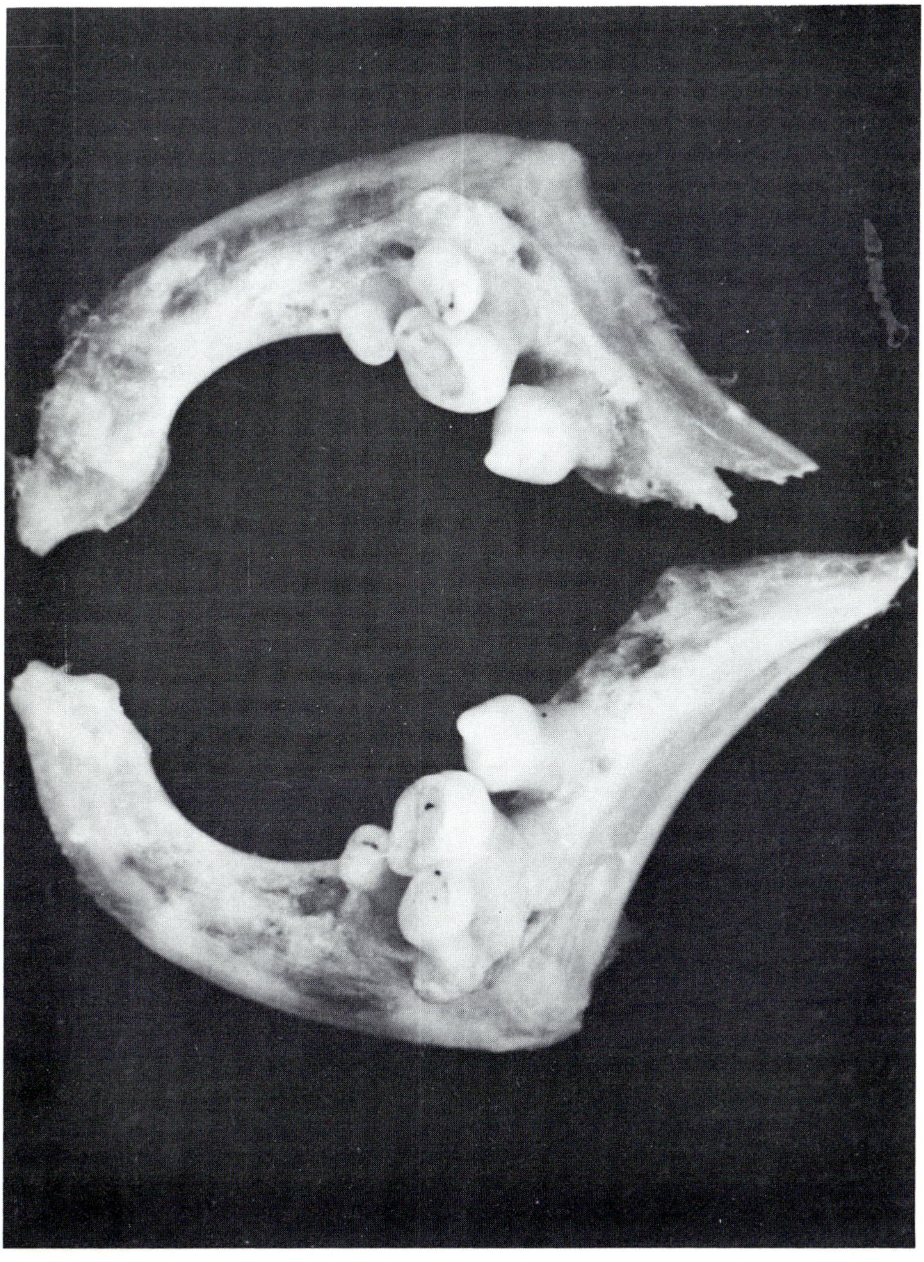

Fig. 2 Pharyngeal teeth of dwarfed wild carp no. 6, six-year-old female; tooth formula 1.3 – 3.1, approx. x8.

TABLE 2. Morphometric and meristic characters of wild carp (*Cyprinus carpio carpio*) from the Danube River and of its "dwarfed" offspring from aquaria

*Meas-ure-ments**	*Characters defined according to Mišik, 1958*	*River N = 100* $\bar{x}$	*(range)*	*Aquaria N = 7* $\bar{x}$	*(range)*
(1–2)	% OF STANDARD LENGTH:				
(1–3)	Length of head	24	(21–27)	30	(29–33)
(1–6)	Length of snout	9	(7–11)	10	(9–10)
(A–B)	Length of barbel I	1.7	(0.3–3.1)	4.2	(3.3–5.2
(C–D)	Length of barbel II	3.6	(1.5–5.6)	4.3	(3.2–5.6)
(6–11)	Length of orbit	3.3	(1.9–4.7)	7.4	(6.7–9.0)
(9–7)	Interorbital width	10	(8–12)	10	(9–12)
(11–3)	Postorbital length of head	12	(10–14)	14	(14–15)
(4–5)	Depth of head	20	(17–23)	22	(19–25)
(1–12)	Predorsal length	45	(41–49)	50	(48–55)
(1–13)	Preventral length	45	(41–48)	50	(50–52)
(1–14)	Preanal length	74	(70–77)	75	(73–76)
(12–15)	Depth of body	28	(23–32)	30	(22–35)
(12–26)	Width of body	17	(11–23)	15	(13–18)
(16–2)	Length of caudal peduncle	20	(17–23)	18	(17–20)
(18–19)	Minimum depth of body	12	(11–14)	12	(9–14)
(20–13)	Pectoral origin to pelvic base	23	(20–26)	22	(20–24)
(13–14)	Pelvic base to anal base	30	(26–33)	26	(26–28)
(12–21)	Length of base of dorsal fin	40	(36–45)	36	(34–38)
(14–16)	Length of base of anal fin	9	(7–11)	9	(7–10)
(22–23)	Length of upper lobe of caudal fin	24	(20–28)	30	(24–37)
(24–25)	Length of lower lobe of caudal fin	24	(20–28)	30	(25–38)
(20–26)	Length of longest pectoral ray	19	(15–22)	21	(19–25)
(13–27)	Length of longest pelvic ray	17	(14–20)	20	(18–22)
(12–28)	Length of longest dorsal fin ray	17	(13–20)	20	(18–22)
(14–29)	Length of longest anal fin ray	16	(14–18)	19	(18–20)
(1–3)	% OF LENGTH OF HEAD:				
(1–6)	Length of snout	39	(35–43)	33	(32–35)
(A–B)	Length of barbel I	7	(3–12)	14	(11–18)
(C–D)	Length of barbel II	15	(8–22)	15	(11–19)
(10)	Internasal width	23	(19–26)	20	(19–22)
(6–11)	Length of orbit	14	(12–17)	24	(22–28)
(7–9)	Interorbital width	42	(38–47)	34	(32–37)
(11–3)	Postorbital length of head	51	(47–54)	46	(45–49)
(4–5)	Depth of head	84	(74–95)	73	(66–79)
(4)	Width of head	63	(53–73)	58	(53–63)
(2–16)	% OF LENGTH OF CAUDAL PEDUNCLE:				
(16–17)	Depth of caudal peduncle	72	(62–82)	82	(53–101)
(17)	Width of caudal peduncle	41	(30–52)	33	(24–46)
(18–19)	Minimum depth of body	61	(51–72)	67	(51–80)
(13–20)	% OF PECTORAL ORIGIN TO PELVIC BASE:				
(20–26)	Length of longest pectoral ray	81	(62–101)	98	(86–110)
(13–14)	% OF PELVIC BASE TO ANAL BASE:				
(13–27)	Length of longest pelvic ray	57	(44–70)	77	(67–84)

Table 2 (continued)

*Meas-ure-ments**	*Characters defined according to Mišik, 1958*	*River* *N = 100* $\overline{x}$ *(range)*	*Aquaria* *N = 7* $\overline{x}$ *(range)*
	FIN RAYS:		
(D)	dorsal	(II) III–IV 18–21 (22)	III (18–19) 20
(A)	anal	(II) III 4–5	III (4) 5
(C)	caudal	IV–VIII 16–18 IV–VIII	VI–VII 17 (18) VI–VIII
(P)	pectoral	I (14) 15–18 (19)	I (15) 16 (17)
(V)	pelvic	II 7–9	I 8 (9)
(K) left	Total number of scales	(34–36) 37–39 (40)	(35) 36, 38
right	in lateral series	(35–36) 37–39 (40)	(35) 36 (37)
above	Number of scales in	5–7	5 (6)
below	transverse series	5–7	5 (6)
(I) outside	Number of gill-rakers	(22) 23–27 (28)	(22) 24, 26 (27)
inside	on the first gill arch	(29) 30–34 (36)	29 (31)
(J)	Pharyngeal teeth	1.1.3–3.1.1	1.2.1–3.1|1.3–3.1| 1.2.1–1.2.1

* See figures on inside front cover.

nor domesticated pond carp had so many irregular teeth as experimental carp (Fig. 2), though domestic pond carp show similar irregularities in 15% more cases (Steffens, 1964) than wild carp. Dwarfed carp had considerably longer heads, a longer first pair of barbels, larger eyes, longer preventral distance, longer caudal fin lobes, and ventral fin rays (Table 2) than wild carp from natural habitats. Dwarf carp were also characterized by a slightly larger postorbital length of head, greater depth of head, greater predorsal length, greater depth of body, greater length of the longest pectoral fin, and longer ventral, dorsal, and anal fin rays in relation to the standard length. No characters are considerably smaller in the dwarf fish except interorbital width and postorbital length in relation to length of head. Length of snout, depth and width of head, and length of the dorsal fin base are somewhat smaller than those measurements in wild fish.

Similar differences in body proportions between wild and domestic carp were found by Steffens (1964). Similar changes in morphometric characters were described by Wunder (1949) for starved carp and for starved European bream (*Abramis brama*) by Lühmann and Mann (1957). In contrast, Rudziński (1961) reported that wild Danube carp in ponds had a smaller head than a domestic carp of the type he studied.

The size of the gape of the mouth, a character that Rudziński (1961) and Steffens (1964) used so successfully to distinguish wild and domestic carps, was considerably smaller in the dwarfed fish than in the domestic or wild carp (Table 3). Rudziński and Steffens considered the enlargement of the mouth in domesticated animals to be an adaptation to changes in feeding habits and also a probable result of artificial (man-induced) selectivity. Domestic carp selectively adapted to utilize supplementary food added to ponds grew better in ponds when artificial food was added. Is it possible that the small amount of food given in small particles to dwarf carp of my

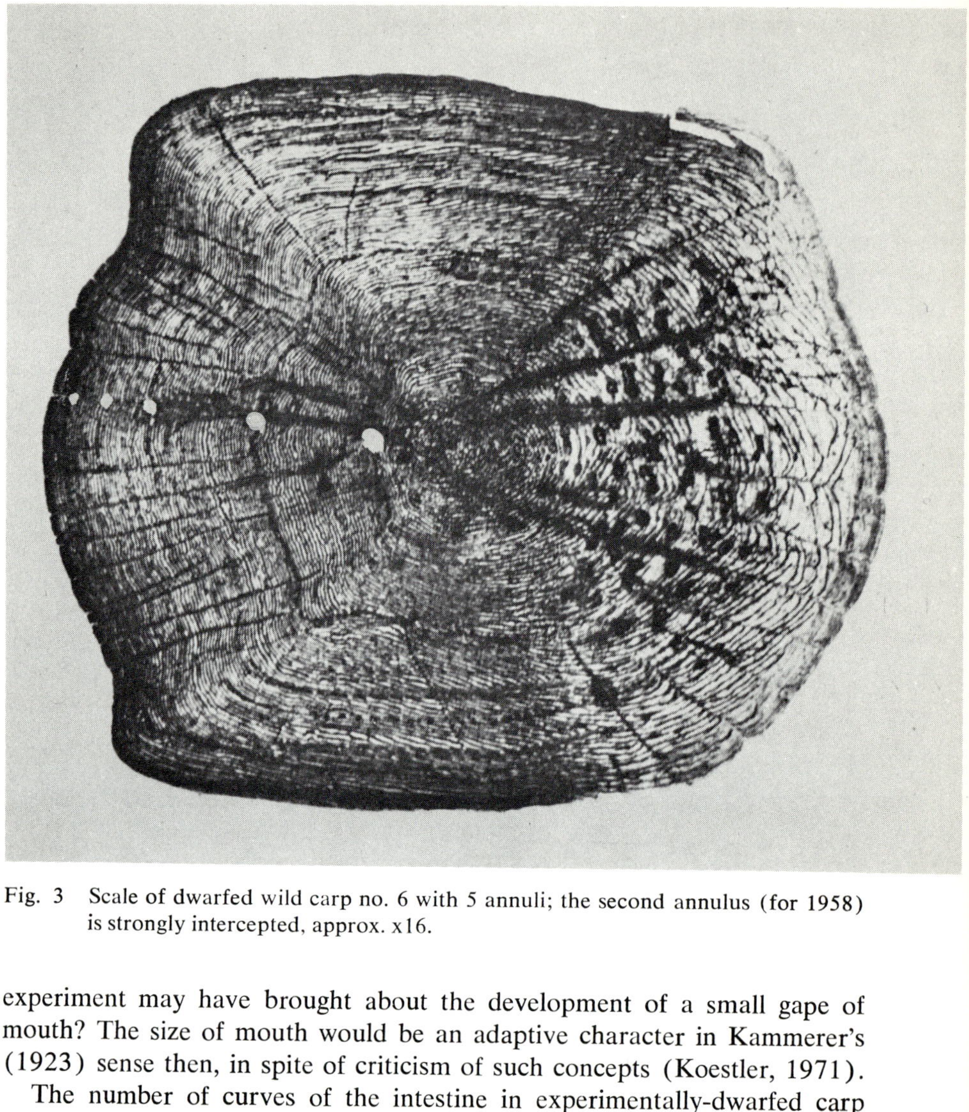

Fig. 3 Scale of dwarfed wild carp no. 6 with 5 annuli; the second annulus (for 1958) is strongly intercepted, approx. x16.

experiment may have brought about the development of a small gape of mouth? The size of mouth would be an adaptive character in Kammerer's (1923) sense then, in spite of criticism of such concepts (Koestler, 1971).

The number of curves of the intestine in experimentally-dwarfed carp varied (4,7,4,4,5,6, respectively) as did other characters (Table 3). Adult domestic carp average six curves to the intestine (Klust, 1939). Hence not all dwarf fish developed fewer curves. Rudziński (1961) found, and Steffens (1964) confirmed, that the intestine of wild carp was generally 15–25% shorter than that of domestic carp. Artificially "dwarfed" carp developed the shortest intestine which was less than 50% the length of that of the domestic carp and slightly more than 50% of that of the wild carp. Differences are even more emphasized if related to body weight: the ratio of gut length to body weight is 2.2 in domestic carp, 3.0 in wild carp, and 8.3 in dwarfed carp (Table 3).

In spite of experimental conditions of minimal temperature fluctuation, the absence of seasonal changes, and constant feeding throughout the year,

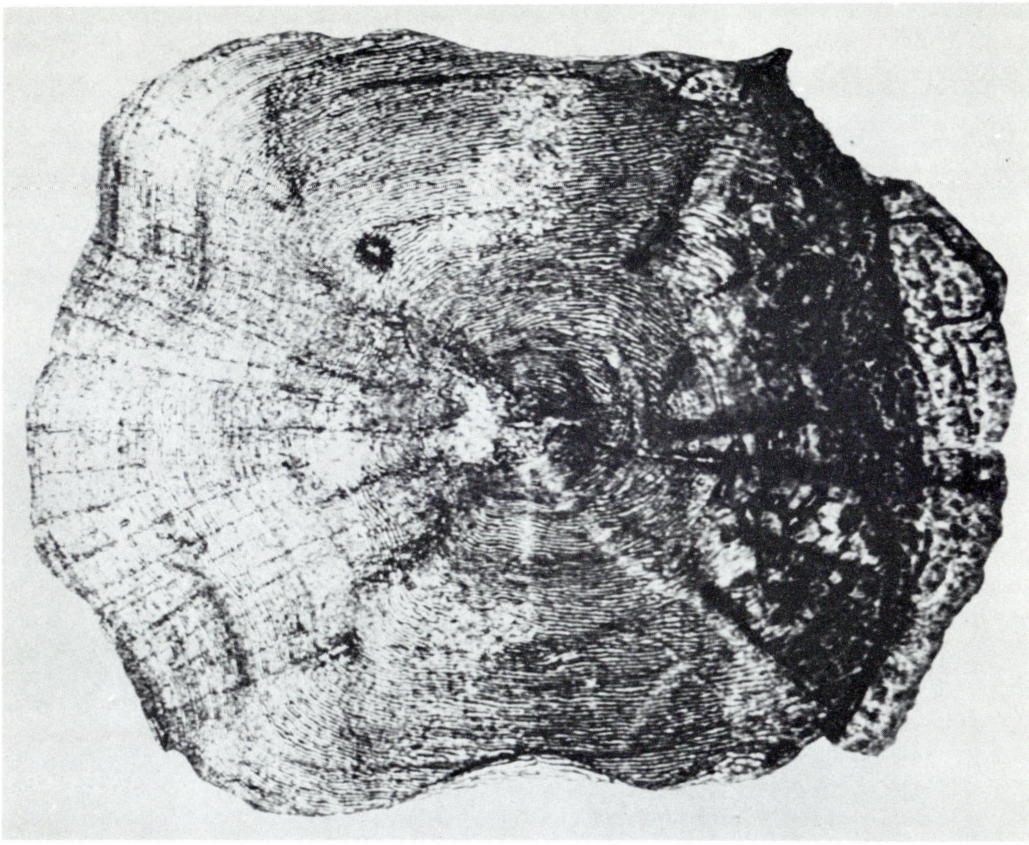

Fig. 4 Scale of dwarfed carp no. 7, with absorbed dorsal and ventral edges, approx. x18.

scales of dwarfed carp developed regular annual marks (Fig. 3). The accidental drop in temperature in 1958 was indicated in the scales by a pronounced annulus. The shape of most key scales was regular but heavy absorption at the dorsal and ventral edges (Fig. 4) occurred in some collected toward the front or rear end of the body during the last two years. The first key scales were collected from the series above the lateral line, dorsal to the origin of the pelvic fin. True annuli were readily distinguished, but between them there were several confusing false annuli. I was, however, unable to relate scale growth to body growth, and values of body size back-calculated from scales for previous growth seasons do not make sense. Interception of annuli may be initiated by the intervals of life history and "the planetary yearly system" (Balon, 1971b, p. 99), but temperature or other environmental factors may strengthen the annulus interception (Segerstråle, 1932).

The above data are recorded as a stimulus to interest, and in order to complete the description of the artificially dwarfed carp. The results of this

TABLE 3. Mouth-gape and intestine lengths indices for dwarfed, domestic and wild carp: o/l = 10 × mouth-gape (cm^2)/standard length (cm), o/w = 10 × mouth-gape (cm^2)/weight (g), o/lc = 10 × mouth-gape (cm^2)/length of head (cm); gut/l = length of intestine (cm)/1 cm of standard length, gut/w = length of intestine (cm)/10 g of weight.

Index		*o/l*	*o/w*	*o/lc*	*gut/l*	*gut/w*
	1	0.10	0.19	0.32	1.14	21.22
	2	0.38	0.47	1.17	1.49	18.09
Dwarf wild carp	3	0.49	0.11	1.65	0.67	1.54
(aquaria)	4	0.47	0.10	1.58	0.84	1.80
	6	0.37	0.12	1.28	1.10	3.48
	7	0.54	0.12	1.78	1.77	3.87
	$\bar{x}$	0.40	0.18	1.30	1.17	8.33
Domestic carp						
(Rudziński, 1961)		1.91				
(Steffens, 1964)						
supplementary diet		2.00	0.08	8.12	2.64	2.25
natural diet		2.30	0.08	8.12		
Wild carp						
(Rudziński, 1961)		1.14				
(Steffens, 1964)						
supplementary diet		1.09	0.10	4.46	2.11	3.02
natural diet		1.24	0.08	5.27		

experiment extended conclusions of an earlier study (Balon, 1960b) beyond the early intervals of fish life. Karzinkin's (1935) experiments suggested that piscivorous fishes would be less capable of survival on substitute foods, although some species, at least, produce dwarfed morphs (Popova, 1967) if population survival is endangered. Some species that forage, however, may not only survive when starved but if later given an optimum diet overtake well-fed individuals (Křiženecky and Křiženecká-Pulanková, 1953; Kuznecov, 1957). Great viability is a characteristic of wild carp, which tolerate a wide range of environmental factors and are consequently a unique teleost species—a most successful colonizer of the world.

Probable Origin of Domestication

The most western natural occurrence of rheophilic wild carp in the Danube is at the mouth of the Morava (March) River. It was there, near Devin (Töben), that the well-known Amber Road crossed the Danube River. The Celtic settlements on the Danube River and the lower parts of its tributaries in southeastern Slovakia "existed until the middle of the first century A.D. and sometimes earlier (compare the settlement at Devin in the second half of the first century), for the Latin material culture often mixed with the Dacian" (translated from Pelikan, 1960). The carp was known to the Romans, who travelled along the Amber Road through territory where huge schools of wild carp probably spawned every year on flooded meadows. Both facts indicate that the Danube and Amber Road intersection is the site from which transfer of the carp to European waters outside the Danube area probably began. Other places are less probable, even though some were suggested. Movtchan (1966) even assumed that Romans transported carp to Italy from Fanagoria (the area of Kuban) and maintained them in ponds. Possibly, again, Dubravius (1547) was confused by the transfer of carp to brackish ponds and by its subsequent multiple spawning, which is, however, usual for carp in warm water (Buschkiel, 1932). Conseqently, Dubravius' interpretation of Pliny's (Plinius, A.D. 24–79; 1635) data concerning multiple spawning of carp in Italy was incorrect and distracted his attention from the true geographic origin of domestic carp.

ROMAN GOURMETS AND FIRST CARP IN CAPTIVITY

Although fish were captured in the Danube 10,000 years ago by the Cromagnons of Upper Paleolite (France), the first proof of fishing activity came from excavations of late Neolithic Stone Age (Clark, 1948). Childe (1929) assumed that "... important are the deer's horn harpoons from Vinča and Csóka (fig. 17), for they show how much the inhabitants of these sites appreciated the good fishing of the Danube." When the Romans arrived they encountered a well-developed fishing technique which was sufficiently documented in the first writings on fishing in the Danube River, scattered through the works of Strabo (63 B.C.; 1917–1932), Plinius (A.D. 24–79; 1635), Aelianus (A.D. 170–235; 1858), and Ausonius (A.D. 310–393; 1933). About A.D. 15 the Romans built a camp opposite the mouth of the Morava River on the site of a Celtic village that later became their biggest Pannonic town, Carnuntum, an important resting place on the Amber Road. Stone reliefs from that period frequently used the fish motif, which I had in two cases identified as a carp (e.g., Fig. 5, Komarno City Museum). In the first century A.D. the XIVth and XVth Roman legions were stationed on the site of the present city of Bratislava, then called Peiso-Piso. Other garrisons and troops, living at sites along nearly the entire length of the Danube River, certainly fished for their livelihood as can be gathered from written remarks made by Roman authors spoken of earlier, as well as from works of Herod-

Fig. 5 Stone relief with carp, partly spoiled by later sculptured head of a goat, from the Roman excavations at Komarom (Komarno City Museum). Photo E. K. Balon, 1966.

otus (485–425 B.C.; 1920–1924), Paterculus Velleius (19 B.C.–A.D. 31; 1924), Tacitus (A.D. 55; 1836), and Cassiodorus (A.D. 490–585; 1626). The commanders of these garrisons were members of patrician families and as such were brought up on the Roman tradition of epicurianism. It must, therefore, have been a source of pride to bring back to Rome an as yet unknown delicacy. In Pliny's Historiae Naturalis, for example, are found such remarks as:

Et in Danubio Mario extrahitur, porculo marino similimus . . . ,	And in the Danube River is fished hausen (*Huso huso*) resembling a sea dolphin . . . ,

and in Aelianus (loc. cit.) we find detailed descriptions of fishing through ice. As early as the first century B.C. Cicero's teacher of gastronomy, Sergius Orata, had devised special salt water reservoirs separated from the sea where he stored fish for the kitchen. These reservoirs ensured a permanent supply without regard to weather and success in fishing. According to Plinius (A.D. 24–79; 1635) this method was adopted by Lucinius Muraena, who began storing freshwater fish. The patricians liked this manner of keeping their fish and competed in establishing such "piscinae," often spending enormous sums of money on them. Consul Lucullus (75 B.C.), whose reputation as a gourmet is well known, dug through a hill near Naples to bring water to his ponds, which were reputedly more costly than his villa.

Although the original idea of the *piscinae*, as devised by Sergius Orata, was to store fish for gastronomic purposes, rearing fish later became a hobby. Another patrician, Hortensius, became as famous as Lucullus, but for the love of his ponds (Hortensius liked his eel so much that he wept when he found it dead . . .), and was even accused by Cicero of neglecting politics because of his fish. The Roman preferred sea fishes, as Varro (116–27 B.C.; 1912) emphasized; freshwater ponds were apparently considered inferior and plebeian, but documented prejudice is at least proof of the existence of freshwater ponds.

Now perhaps it can be understood why visiting senators, patricians, and plebeian soldiers not only tasted the fish during their temporary stay in the Danube region but tried to transfer them alive to *piscinae* at home, either as fish fanciers or as gourmet attractions for guests. Which of the largest, tastiest fishes of this region would have survived the rigours of primitive transport? Which could have lived in saline or brackish water after such transport? The wild carp is among the least sensitive of fish and can tolerate water with a low oxygen content so that it may be easily transported and afterwards be kept for weeks in small reservoirs (transportation in wet moss or other moisture-retaining materials is probable). Moreover, no special food is needed and it can endure long periods of starvation. Besides having palatable meat, carp has one very important ability—it lives in waters with salinities as great as 6,000 mg Cl/l (Nakamura, 1948; Johnson, 1954; Mark, 1966). Wild carp even live in the brackish waters of the delta of the Danube (Bănărescu, 1964).

The above, then, is how I envisage the Danube wild carp to have been transferred to Italy at the beginning of our era to begin life in *piscinae*. From there some escaped to local rivers and, after the collapse of the Empire and the establishment of Christianity, were reared in monastery ponds. I do not exclude the possibility that wild carp endemic to the Danube River were later brought to other west European ponds. Apparently the messages of Roman and Middle Age scholars describing these early imports were passed on from generation to generation and probably accounted for the reestablishment of the carp-rearing tradition. In support of the above let me quote Cassiodorus (A.D. 490–585; 1622):

. . . destinet carpam Danuvius: A Rheno veniat anchorago (. . .): sapori pisces de diversis finibus afferantur. sic decet regem pascere, ut a legatis gentium credatur paene omnia possidere.

. . . from the Danube come carp and from the Rhine herring. To provide a variety of flavours, it is necessary to have many fish from many countries. A king's reign should be such as to indicate that he possesses everything.

Apparently King Theodorus (A.D. 490–526) of Ravenna (Italy) ordered the transport of carp from the Danube to his country, thus imitating Roman tastes and continuing the still novel introductions.

It was only 100 years ago that Dubisch brought Danube wild carp to the Upper Vistula River for hybridization with the local form of domestic carp (see Morcinek, 1909). In 1957 I repeated Dubisch's experiment by transferring 1,000 yearling Danube wild carp to the ponds at Ochaby (experimental station of the Polish Academy of Sciences) (Rudzińsky, 1961). But as in experiments on the Danube, studies at the Ochaby pond proved (Leszczyńska and Biniakowski, 1967) that wild carp are more suitable for stocking into natural waters than are domestic carp.

FASTING MONKS REDISCOVER THE CARP

Christianity introduced more than 100 fasting days a year, and the only meats that could be eaten during those days were crayfish, molluscs, and other cold-blooded animals such as fish (in some regions fowl and/or unborn rabbit embryos, *laurices*, were exceptions). As heavy punishment, which sometimes went as far as the death penalty, was meted for violations of proscriptions, a readily available supply of fish was important. Thus, fish became the only possible foods for monks during a substantial part of the year. Although fish were abundant in rivers and lakes, there were many days in the year when weather conditions, natural disasters, or wars made fishing impossible. Monks and missionaries, who had to follow fasting regulations, had difficulties finding fish (Leonhardt, 1906). Probably after monks settled and established a monastery, they thought about keeping a fish supply as they had in their southern homeland; they also remembered or read how suitable the carp was. The first monasteries were founded in the early 6th century (e.g., Monte Cassino Monastery was founded A.D. 529). Later they gained land and farms, and conditions became favourable for the beginning of mass culture of fish that provided the monastery a ready supply of food for fast days. Monks, according to Leonhardt (1906), first reared local fishes; inland monasteries

Fig. 6 This illustration from the 18th century shows that monks, very much like the Romans, enjoyed fishing for carp in monastery ponds. (Courtesy of the Mansell Collection, London.)

kept pike (*Esox lucius*), crucian carp (*Carassius carassius*), bream (*Abramis brama*), and other species. But these species were difficult to keep in good condition in the primitive reservoirs. They knew that carp could be bred easily in these circumstances and introduced it.

REARING OF CARP IN PONDS AND FIRST REPRODUCTION IN CAPTIVITY

Carp then were first reared with other species of fish in a simple man-made rearing pond. Certainly, some unexpected spawning occurred already in Roman *piscinae* and also in monastery ponds; no organized reproduction, however, was recorded. Charlemagne (A.D. 768–814), the first Holy Roman Emperor, ordered his tenant farmers to maintain ponds and issued orders for their control. His orders, however, were concerned with protection against poaching, regulation of fishing and sale of fish, not with culture. Leonhardt (1906) claims that the lack of concern in carp reproduction was conditioned by an abundant fish fauna in local waters with which the landlords could regularly stock the ponds. In my opinion, Leonhardt (loc. cit.) incorrectly assumes the natural occurrence of the wild carp then in waters of southern and northern Europe. Moreover, archaeological findings of Slavonic settlements at the outset of the Polish Empire in the Ninth to Twelfth Centuries did not include carp, though they have produced remains of ide (*Leuciscus idus*), sturgeon (*Acipenser sturio*), chub (*Leuciscus cephalus*), tench (*Tinca tinca*), perch (*Perca fluviatilis*), mud loach (*Misgurnus fossilis*), roach (*Rutilus rutilus*), beaked carp (*Chondrostoma nasus*), wels (*Silurus glanis*), pike (*Esox lucius*), eel (*Anguilla anguilla*), salmon (*Salmo salar* or *Salmo trutta*) and rapfen (*Aspius aspius*) (Perlbach, 1881). The first written evidence of carp in Polish territory is dated 1466 and originates from the area of Kołomyje in the Black Sea Basin (Chmielewski, 1965). Later carp were kept in ponds as proved by the following words of M. Rej (A.D. 1505–1569) quoted by Górzynski (1964):

. . . a wszystko to bardzo małem zachodem otrzymać można: bo siedem, ale dziewięć karpi puścić, takież w drugą (sadzawkę) karasków, ali ty i pieniążki i pożytek z tego mieć możesz.

. . . and it is possible to obtain everything without trouble: release seven or nine carp, then the same number into another pond of crucian carps, and you will benefit from it as well as have money.

But the pond culture of carp still cannot be connected with its pond reproduction.

Hildegarde (1089–1170), mother superior of the Bavarian Convent of the Benedictine Order, mentioned in her recipes the preparation of carp (Koch, 1925); Albertus Magnus (1193?–1280; 1861) first wrote about breeding it in ponds. But generally not much information is available about the rearing of carp in the Middle Ages. It is as if the secret had been kept within the monastery walls. Gradually, however, complete articles, even simple studies about the rearing of the carp and its culture in ponds appeared (Dubravius, 1547; Strumieński, 1573; Strojnowski, 1609; von Hohberg, 1687), the reinstatement of rearing in the late Middle Ages is well known (Leonhardt, 1906; Steffens, 1958). Von Hohberg (1687) considered that the rearing methods of his time were superior to those used by the Romans and also that the Romans would surely have enjoyed the taste of the real contemporary carp. In the work of Dubravius (1547) numerous remarks about the beginning of Roman domestication of carp piece together the historical connections.

ONE MORE ORIGIN OF DOMESTICATION

Domestication of wild carp in China began independently, occurred five centuries earlier than in Europe (Leonhardt, 1906; Tamura, 1961; Hickling, 1962), and probably involved the local subspecies *Cyprinus carpio haematopterus*. The latest, anonymous Chinese study (the following quotations are from the Russian translation—Anonymous, 1961) on pond culture stated that "thanks to the creative efforts of the Chinese people for many generations, breeding of the carp in this country has proceeded successfully for more than 2,000 years. From China the breeding of this fish spread all over the world." Further, it was Fan Lio (Tao Tschshugun) who succeeded in spawning the carp and in growing the young to adulthood in a pond at the time of the "Spring and Autumn" and the "Fighting Dynasties" (in the eighth–third centuries B.C.). "From Asia the rearing of carp spread to Europe and later to America, Australia, and Africa." There is no reason for the Chinese today to contradict an opinion that supports their national pride, particularly as most European authors supported the Chinese origin of the carp's domestication. The Chinese are proud of their past isolation and independent cultural development. As China was virtually inaccessible until a century ago, would it not be improbable for the carp to have been brought from there in the first years of the Gregorian calendar or the beginning of the Middle Ages? An independent, second introduction of carp to Western

Europe probably occurred during the Middle Ages and is considered by some to have been from China (Rudziński, 1962). Previously mentioned evidence favours the Danube as a source also of the later introductions. As a result of the introduction into Asia of European pond carp, however, the carp in ponds in some regions in eastern Asia may be mixed in origin (Buschkiel, 1933).

Consequences of Domestication

As early as the sixteenth century, races of carp were recognized, especially those morphs in which scales were few or absent. The domestic carp of western Europe was introduced into North America in 1831 for the first time, again in 1872, and frequently later (Hessel, 1881; Bartlett, 1901, 1905; Cole, 1905; Dymond, 1955; Atton, 1959; McCrimmon, 1968). The carp was introduced into Australia in 1860 (Butcher, 1962, 1967), and in 1896 to the Cape of Africa (Jubb, 1967). According to my hypothesis all these carp originated from the rheophilic wild carp of the Danube River. While this hypothesis is geographically and historically attractive, in the absence of more evidence it remains still speculative. With this in mind, the gold carp (*Cyprinus auratus*) of the Schuykill and Massachusetts rivers (Forester, 1850, quoted by McCrimmon, 1968) could be a scaled carp whose genotype was predominantly the wild morph and so selected a more lotic habitat than other carp in the shipment whose genes were primarily of more advanced domestic carp. (The most freshly caught Danubian wild carp that I saw were always a clear golden colour on the scaled region of their bodies.)

In Europe the domestic carp became the most widely cultured and consumed fish, and pond culture gradually became one of the sophisticated branches of agriculture. The same species in North America, however, is considered to be undesirable, probably because of its unpalatable meat as compared to that of many local fishes and because of an adverse interaction with the aquatic habitat. Even in Australia "The European carp (*Cyprinus carpio*) has been proclaimed by legislation to be a 'noxious fish' and the keeping and/or release of this species is prohibited" (Butcher, 1967).

Consequences of domestication of wild carp become apparent when the wild morph from the Danube is compared with any established domestic morph. Subsequent to rediscovery of rheophilic wild carp in the Czechoslovak–Hungarian section of the Danube (Balon and Mišik, 1956) and initial studies on it (Balon, 1957, 1958a; Mišik, 1958), morphometric and growth indices were compared with those of the domestic populations (Tuča, 1958; Steffens, 1959; Bastl, 1961, 1962; Rudziński, 1961; Chytra et al., 1961; and Mišik and Tuča, 1965). Some morphological differences between wild and domestic carp were stressed by Rudziński (1961) and Steffens (1964). It will be sufficient to mention some of the results of the latter study:

The body of the domestic carp, which is nearly always much deeper than that of the oblong, cylindrical body of the wild carp, appears to have more flesh, but the calculated ratio of muscle in both morphs is the same. Without regarding changes in proportions, dressed weight of individual domesticated carp did not increase even though its greater growth rate produces more absolute meat in a given period of time. Probably the faster growth of domestic carp can be correlated with the larger mouth and longer intestines acquired by the pond form following its adaptation to complementary nutrition and perhaps also by man's selection.

The greater strength, mobility, and viability of wild carp are emphasized by physiological characteristics. The wild carp has 18% to 19% more erythrocytes and haemoglobin than does the domestic carp. Blood sugar level is 16%–26% higher. The wild carp has a much lower water content in muscles and liver than does the domestic carp. Furthermore the wild carp has a greater content of fat in individual organs, of glycogen in the liver, and of vitamin A in the intestines, eyes, and liver. Consequently the taste of the wild carp is better because the flesh is juicier. The wild carp is more mobile, stronger, and nimbler because its muscles are better supplied with blood ensuring a better supply of nutrients and elimination of waste; muscles do not fatigue as quickly as do those of the domestic carp, enabling wild carp to overcome the river current. The same improves the taste of the flesh of the wild hare and the wild rabbit, said to be more juicy than the domestic rabbit (Volf, 1965).

Epilogue

The impressive number of these that are 'not known in the wild state' emphasizes the drastic nature of the morphological changes that domestication has so often brought about in the transition from wild ancestor to cultivate crop.

David R. Harris (1967, p. 91)

Perhaps the unpalatable taste of carp meat in North America is a result of the carp's origin. Nearly all carp that live in natural waters here are descendants of domestic carp which escaped into natural habitats. Limited ability to utilize natural food and poor growth may have resulted in unpalatable flesh in feral domestic carp. European domestic cattle, when left to become wild, are known to remain in poor condition and never revert to a condition similar to that of the strong, wild, ancestral auroch (*Bos primigenius*) (see Talbot et al., 1965; Taylor, 1970, 1972). Perhaps the carp fall to feral condition parallelled that of the cattle. If this is the case it is not surprising that in North America carp "have never been wholeheartedly accepted by the general public as an item of food" (Crossman, 1969, p. 145).

The need for fasting food may have been the main reason for rediscovery of the carp in the Middle Ages, but not everywhere has the hedonistic approach of the Romans to rearing of the carp given way to pure necessity; there is some evidence (Fig. 6) of monks enjoying fishing for carp (Burton and Burton, 1968). Nevertheless, the need for a larger variety of foods during fasts undoubtedly played an important role in bringing about the tradition of carp eating among some European Christians. Though related to fasting, this tradition was eventually also embraced by some European Jews who made the carp their Sabbath meal.

That the history of the domestication of the carp is generally similar to that of the European rabbit (*Oryctolagus cuniculus*) may be of some interest as peripheral support of my hypothesis on carp origin. Varro (116–27 B.C.; 1912) wrote that Romans brought rabbits from Spain and bred them in special enclosures called *leporaria* to ensure a fresh supply of meat at all times. Unborn rabbits collected from pregnant females were prepared as a special dish called *laurices*. At the beginning of the Middle Ages rabbits were a popular fare in western European monasteries, and *laurices* were eaten even during fasting periods. (It was in the corridors and paved courts of the monasteries that the rabbits began giving birth to their young above ground instead of in burrows. The young gradually became domesticated and accustomed to people.) In the middle of the sixteenth century several colour morphs of rabbits were known. An albino rabbit is shown in Titian's "Madonna" (1530) at the Louvre (Valcanover, 1960). Around 550 B.C. the Chinese philosopher Confucius (1898) suggested to poor farmers that they keep some rodents similar to rabbits (Volf, 1965). Unlike speculation on the origin of the carp, the origin of the domestic rabbit was not assumed to be in China. Otherwise the similarity of the domestic history of both animals is striking.

Wild rabbits are overabundant in many countries of the world and consequently special measures are taken to control them. Conversely, rheophilic wild carp in Europe today is in danger of extinction. Regulation of the flow of rivers and construction of waterworks are destroying remaining spawning areas. In the Danube River, between the mouths of the Morava and Hron Rivers, spawning schools that were plentiful as recently as 15 years ago have greatly decreased. At a decreased population density the wild carp may lose its habitat to the domestic carp, which is repeatedly stocked, purposefully and accidentally, into rivers. Competitive replacement may be accelerated if planned hydroelectric plants and waterways are introduced (Balon, 1967a, 1967b).

Scientifically directed selection would have a better chance of improving cattle production and quality if genes of the ancestral auroch were still available. The wild carp may be compared to the extinct auroch when consequences of its possible extinction are contemplated. In 1965 Slovakia included spawning schools of wild carp in the official list of animals protected by law (Randik, 1967). Even so, because of the rapid deterioration of the environment, extinction is the probable fate of this species. Protection of the original stock of wild carp in the foothills of the Danube River is therefore an important task for mankind. It was probably here that the most useful domestic fish of the world originated.

Summary

If preglacial remnants of the carp from pile dwellings are rejected as incorrectly dated, it is postulated that the wild carp originated in central Asia and spread naturally east into China and adjacent regions and west as far as the Danube River in the last postglacial period. That this did not happen earlier is suggested by the fact that in the Pleistocene, after the retreat of the continental glaciers, *C. carpio* was not among the species which penetrated either Scandinavia, the British Isles, or North America. Had it occurred at that time in eastern Asia it probably would have invaded North America together with the pike and other freshwater fishes via the Bering land bridge.

The natural occurrence of the wild carp in the Danube River is inferred from the different names given to the carp by the local inhabitants and from the use of the same Celtic name outside the western limit of its natural range; further evidence comes from medieval chronicles concerning the occurrence of this fish, and from contemporary studies of the rheophilic wild form of the carp in the Danube River. The ancient differentiation of wild carp in central Asia and presence of undifferentiated isolated single morphs in the eastern and western regions where it occurs suggest again that its place of origin must be sought in central Asia.

A separate section is devoted to an experiment testing the ability of Danube River wild carp to survive in starvation condition; this demonstrates the hardiness of the animal and explains why it was domesticated for human consumption.

Finally, from written notes of the Romans and excavated art objects depicting the carp, the theory is presented that the first carp brought to Roman *piscinae* came from places where the Amber Road—the most travelled thoroughfare of those times—crossed the Danube River. These localities, regularly inundated meadows where schools of carp spawned, occurred at the mouth of the Morava River, near ancient Carnuntum and the later Devin Castle. Patrician commanders of the Roman legions stationed there brought the carp to Rome as an epicurian delight, for it was one of the most hardy species of fish and survived primitive techniques of transport and rearing. After the decline and fall of the Roman Empire, the rearing techniques of carp in Roman *piscinae* were not forgotten. Christian missionaries and monks needed ready reserves of fish for their fasting days. From Latin writings and from traditions of their southern homelands they knew that the carp was appropriate for such purposes and looked after their supply as soon as they settled somewhere. After the Middle Ages breeding had become routine, not only in monasteries but also on private estates.

The history of the domestication of the carp in Europe that began with wild carp from the Danube River (*Cyprinus carpio carpio*) is divided into five periods: 1. First introduction by the Romans west of the Danube River in the first to fourth century A.D.; 2. sporadic western introduction and rearing in the fifth and sixth centuries; 3. beginning of rearing on a mass scale, with some renewed introductions, and initial attempts to breed it in greater quantities in specially-built ponds in the seventh to thirteenth centuries; 4. beginning of breeding, mass culture, and fortuitous selection in the fourteenth to sixteenth centuries; 5. intensification of breeding, purposeful selection, and introductions into North America, Australia, the Far East, and Africa from the seventeenth century. In China domestication of the carp, probably from local subspecies *C.c. haematopterus*, was begun five centuries earlier but independently of similar efforts in Europe.

The most important features by which the wild carp from the Danube River differ from the present domesticated forms are stressed and reference is made to a striking similarity with the domestication of the rabbit. Attention is drawn to the necessity of preserving the last wild schools of this fish in the foothill-zone of the Danube River and of protecting the wild predecessor of the domestic carp and its habitat. Most important, a repetition of the fate of the extinct auroch, the ancestor of cattle, should not be permitted. As with cattle, domestic carp in wild conditions becomes an animal of poor food quality, which is perhaps the reason why the carp deteriorated so after its transatlantic introductions.

Acknowledgments

Above all, I owe an outstanding debt of gratitude to V. Mišik, V. Tuča, I. Bastl (Slovak Agricultural Academy, Bratislava) and E. Rudziński (Polish Academy of Sciences, Cracow) for their original studies of the Danube wild carp that were consistent with my ideas and which led to the formulation of the theorem here presented. I thank J. Holčik (Slovak National Museum, Bratislava), O. Oliva (Charles University, Prague) and A. G. Coche (UNDP/FAO, Chilanga), who pointed out errors or inadequacies in the first draft, and E. J. Crossman (Royal Ontario Museum, Toronto) for reading, correcting and criticizing the final draft.

Literature Cited

AELIANUS, C.

1858 De natura animalium. Varia historia, etc. R. Herrher, Parisiis. 541 pp.

ANONYMOUS

1961 Prudovoye rybovodstvo (Pond fisheries). Izd. Selskochoz. lit., Moscow.

ARISTOTLE

1862 History of animals. In ten books. Translated by Richard Creswell. London, H. G. Bohn. 326 pp.

ATTON, F. M.

1959 The invasion of Manitoba and Saskatchewan by carp. Trans. Am. Fish. Soc., vol. 88, no. 3, pp. 203–205.

AUSONIUS, D. M.

1933 The Mosella. Translated into English verse by E. H. Blakeney. London, Eyre & Spottiswoode, 74 pp.

BALON, E. K.

1957 Vek a rast neresového stáda dunajského kapra – sazana (*Cyprinus carpio* morpha *hungaricus* Heck.) z Malého Dunaja nad Kolárovom [Age and growth of spawning shoal of the Danube wild carp (*Cyprinus carpio* morpha *hungaricus* Heck.) from the smaller Danube above Kolarovo]. Pol'nohospodárstvo, vol. 4, pp. 961–986. [In Slovak, with Russian and English summaries].

1958a Vývoj dunajského kapra (*Cyprinus carpio carpio* L.) v priebehu predlarválnej fázy a larválnej periódy [Development of the Danube carp during the prelarval phase and larval period.] Biol. práce, vol. 4, no. 6, pp. 5–54. [In Slovak, with Russian and German summaries].

1958b Die Entwicklung der Beschuppung des Donau-Wildkarpfens. Zool. Anz., vol. 160, pp. 68–73.

1960a Über die Entwicklungsstufen des Lebens der Fische und ihre Terminologie. Z. Wiss. Zool., vol. 164, pp. 294–314.

1960b Die Entwicklung der Fische bei ungünstigen Nahrungsbedingungen. Acta. Hydrobiol., Kraków, vol. 2, pp. 125–132.

1962 Ökologische Bemerkungen über die Standorten der Donaufische mit einer Beschreibung des Fundes des *Carassius auratus gibelio* (Bloch, 1783) und *Alburnoides bipunctatus* (Bloch, 1782). Věst. Čs. spol. zool., vol. 26, no. 4, pp. 333–351.

1963a K nálezu karase stříbřiteho eurasijského [The find of *Carassius auratus gibelio*]. Živa, vol. 11, pp. 205–206. [In Czech]

1963b Altersstruktur der Populationen und Wachstumsgesetzmässigkeiten der Donaubrachsen (*Abramis brama, A. sapa, A. ballerus*). Sb. Vys. Šk. Technologie vody, Chem.-Technol. Prague, vol. 7, no. 2, pp. 459–542.

1964 On relative indexes for comparison of the growth of fishes. Věst. Čs. spol. zool., vol. 28, pp. 369–379.

1967a Vývoj ichtyofauny Dunaja, jej súčasný stav a pokus o prognózu dalších zmien po výstavbe vodnych diel [Evolution of the Danube fishfauna, its recent state and attempts to forecast further changes after the building of hydro-electric power stations.] Biol. práce, vol. 13, no. 1, pp. 5–121. [In Slovak with Russian, English and German summaries].

1967b Ryby Slovenska (Fishes of Slovakia). Obzor. Bratislava, 418 pp. [In Slovak]

1969 Studies on the wild carp *Cyprinus carpio carpio* Linnaeus, 1758. I. New opinions concerning the origin of the carp. Práce Laboratória rybárstva, vol. 2, pp. 99–120. Mimeographed.

1971a The intervals of early fish development and their terminology (A review and proposals). Věst. Čs. spol. zool., vol. 35, no. 1, pp. 1–8.

BALON, E. K.
1971b Age and growth of *Hydrocynus vittatus* Castelnau, 1861 in Lake Kariba, Sinazongwe area. Fish. Res. Bull. Zambia, vol. 5, pp. 89–118.
1972 Possible fish stock size assessment and available production survey as developed on Lake Kariba. Afr. J. Trop. Hydrobiol. Fish., vol. 2, pp. 45–73.

BALON, E. K. AND V. MIŠIK
1956 Zoznam nových dokladov o výskyte niektorých málo známych alebo nových druhov rýb na Slovensku [New evidence on the occurrence of some lesser known or new species of fish in Slovakia.] Biológia, Bratislava, vol. 11, no. 3, pp. 168–176. [In Slovak, with Russian and German summaries].

BĂNĂRESCU, P.
1960 Einige Fragen zur Herkunft und Verbreitung der Süsswasserfischfauna der europäisch-mediterranen Unterregion. Arch. Hydrobiol., vol. 57, pp. 16–134.
1964 Pisces – Osteichthyes. Fauna Repub. Pop. Rom., vol. 13, pp. 1–962. [In Rumanian]

BARTLETT, S. P.
1901 Discussion on carp. Trans. Am. Fish. Soc., vol. 30, pp. 114–132.
1905 Carp as seen by a friend. Trans Am. Fish. Soc., vol. 34, pp. 207–216.

BASTL, I.
1961 Zmena morfometrických znakov dunajského kapra počas rastu v rybničnom chove [Changes in morphometric characters of Danube carp during growth in ponds]. Final Rep. Lab. Fish. Res. Bratislava. Typescript. [In Slovak]
1962 Rast divého dunajského kapra (*Cyprinus carpio carpio* (L.)) v rybnikoch na Kamennom mlyne [Growth of wild Danube carp in ponds at Kamenny Mlyn]. Biológia, Brastislava, vol. 12, no. 10, pp. 757–770. [In Slovak, with Russian and German summaries].

BELON, P.
1555 La nature et diversité des poissons, anec leurs pourtraits, representez au plus pres du naturel. Paris, C. Estienne. 448 pp.

BERTRAM, C.
1963 Man pressure. Oryx, vol. 7, nos. 2 & 3, pp. 97–101 (Reprinted *in* P. Shepard and D. McKinley: The subversive science. Boston, Houghton Mifflin, 1969, pp. 210–215.

BORZENKO, M. P.
1926 Materialy po biologii sazana (*Cyprinus carpio* Linné) [Notes on the biology of the wild carp]. Izv. Bakin. Ikhtiol. Lab., vol. 2, no. 1, pp. 5–132. [In Russian]

BURMAKIN, E. V.
1956 Ob izmenenijach v morfologii sazana, akklimatizirovannogo v basseine ozera Balkhash [On changes in morphology of wild carp, acclimated in Lake Balchash basin]. Zool. Zh., vol. 35, pp. 1887–1891. [In Russian]

BURNS, J. W.
1966 Carp. *In* Calhoun, A. Inland Fisheries Management. Calif., Dept. Fish. Game. pp. 510–515.

BURTON, M. AND R. BURTON
1968 Carp. *In* Burton, M. and R. Burton, eds. Purnell's encyclopedia of animal life. Vol. 1. London, Published by Purnell for BPC Publishing. pp. 383–384.

BUSCHKIEL, A. L.
1932 Studien über das Wachstum von Fischen in den Tropen, I. Über Karpfen in Indien und ihr Wachstum. Int. Revue Ges. Hydrobiol. Hydrogr., vol. 27, pp. 427–466.
1933 Teichwirtschaftliche Erfahrungen mit Karpfen in den Tropen. Z. Fisch., vol. 31.

BUTCHER, A. D.
1962 The implications of the introduction of European carp into Victoria waters. Fish Wildlife Publ., Victoria, Australia. 73 pp.
1967 A changing aquatic fauna in a changing environment. IUCN Publications, n.s., vol. 9, pp. 197–218.

CASSIODORUS, SENATOR, F. M. A.
1626 M. Aurelij Cassiodori Senatoris v.c. opera omnia quae extant, ex fide manuscr. auctiora & locupletiora, . . . P. & I. Chouet, Aureliae Allobrogorum. 1622 pp.

CHILDE, V. G.
1929 The Danube in prehistory. Oxford, Claredon Press. 479 pp.

CHMIELEWSKI, S.
1965 From the history of fresh-water fisheries in Poland. *In* Fresh-water fisheries of Poland. Cracow, Hydrobiological Committee PASc., pp. 5–13.

CHYTRA, F., V. KRUPAUER AND J. PICHA
1961 Růst dunajského kapra v rybnicích v průběhu prvnich tří roků [Growth of Danube wild carp in ponds in the first three years of life]. Prace ČSAZV Vyskumného ustavu hybařskeho ve Vodňanech, vol. 1, pp. 73–91. [In Czech]

CLARK, J. G. D.
1948 The development of fishing in prehistoric Europe. Antiquarics Journal, vol. 28, nos. 1,2, pp. 45–85.

COLE, L. J.
1905 The status of the carp in America. Trans. Am. Fish. Soc., vol. 34, pp. 201–207.

CONFUCIUS
1898 SSU SHU (The Four Books). Hong Kong, Man Yu Tong. 617 pp.

CROSSMAN, E. J.
1969 The Canadian freshwater fish fauna. Ac. Rer. Natur. Mus. Nat. Slov. Bratislava, vol. 15, no. 2, pp. 137–152.

CROSSMAN, E. J. AND C. R. HARINGTON
1970 Pleistocene pike, *Esox lucius*, and *Esox* sp., from the Yukon Territory and Ontario. Can. J. Earth Sci., vol. 7, pp. 1130–1138.

DŁUGOSZ, J.
1863–1887 Chorographia Regni Poloniae. Historia Poloniae. Opera omnia, Cracoviae.

DOLZHENKO, M. P.
1953 K biologii akklimatizirovannogo sazana v ozerakh zapadnoĭ Sibiri [On biology of acclimated wild carp in lakes of west Siberia]. Zool. Zh., vol. 32, no. 6, pp. 1217–1221. [In Russian]

DUBRAVIUS, J.
1547 De piscinis ad Antonium Fuggerum. Rev. ed. *In* Sbornik filologicky ČSAV, vol. 1, pp. 11–75. [In Latin and Czech]

DYMOND, J. R.
1955 The introduction of foreign fishes in Canada. Proc. Int. Ass. theor. appl. Limnol., vol. 12, pp. 543–553.

EPSTEIN, H.
1972 Domestication features in animals as functions. *In* Smith, R. L. The ecology of man: An ecosystem approach. New York, Harper & Row, pp. 91–101.

FOREL, F. A.
1904 Le Léman. Monographie limnologique. Vol. 3. Lausanne, F. Rouge.

GÓRZYŃSKI, S.

1964 Przegląd historii rybactwa w dawnej Polsce [Outline of the history of fishing in ancient Poland]. Warzawa, PWRiL. [In Polish]

HANKÓ, B.

1932 Ursprung und Verbreitung der Fischfauna Ungarns. Arch. Hydrobiol., vol. 23, pp. 520–556.

HARRIS, D. R.

1967 New light on plant domestication and the origins of agriculture: A review. Geogrl. Rev., vol. 57, no. 1, pp. 90–107. Reprinted *in* R. L. Smith, 1972. The ecology of man: An ecosystem approach. New York, Harper & Row, pp. 73–87.

HENSEL, K.

1971 Some notes on the systematic status of *Carassius auratus gibelio* (Bloch, 1782) with further record on this fish from the Danube River in Czechoslovakia. Věst. Čs. spol. zool., vol. 35, pp. 186–198.

HERODOTUS

1921–1924 Herodotus, with an English translation by A. D. Godley. 4 vols. London, W. Heinemann.

HESSEL, R.

1881 The carp and its culture in rivers and lakes; and its introduction into America. Rep. U.S. Commnr. Fish. 1875–1876, pt. 4, pp. 865–900.

HICKLING, C. F.

1962 Fish culture. London, Faber and Faber. 295 pp.

HOHBERG, W. H. VON

1687 Georgica Curiosa oder Adeliches Land und Feld Leben. II Theil. Nürnberg, Michael & Johann Friederich Endters Seel. Erben.

HOLČIK, J. AND K. HENSEL

1972 Ichtyologická príručka [Ichthyological Handbook]. Bratislava, Obzor. 220 pp. [In Slovak]

JOHNSON, M. C.

1954 Preliminary experiments on fish culture in brackish-water ponds. Progve. Fish Cult., vol. 16, no. 3, pp. 131–133.

JORDAN, D. S. AND B. W. EVERMANN

1896–1900 The fishes of North and Middle America. Bull. U.S. Natn. Mus., no. 47, pp. 1–3313.

1902 American food and game fishes. A popular account of all the species found in America north of the equator, with keys for ready identification, life histories and methods of capture. New York, Doubleday, Page. 573 pp.

JUBB, R. A.

1967 Freshwater fishes of Southern Africa. Capetown, A. A. Balkema. 248 pp.

KAMMERER, P.

1923 Breeding experiments on the inheritance of acquired characters. Nature, vol. 111, no. 2793, pp. 637–640.

KARZINKIN, G. S.

1935 K poznaniyu rybnoy produktivnosti vodojemov. Soobshtchenye IV. Usvoenie i prodolzhitelnost prochozhdenia yestestvennych kormov u malkov *Esox lucius* [Toward a wider knowledge of fish productivity in reservoirs. Issue IV. Acceptance and digestion rate of natural foods of pike juveniles]. Trudy Limnol. Sta. Kosine, vol. 20, pp. 81–97. [In Russian, with German summary].

KHIN, A.
1930 Vyzy no Vel'kom Žitnom ostrove a ich lovenie [Hausen on Great Shutt Island and its fishing]. Čas. Mus. slov. spol., vol. 22, pp. 110–112. [In Slovak]

KLUST, G.
1939 Über Entwicklung, Bau und Funktion des Darmes beim Karpfen (*Cyprinus carpio* L.). Int. Revue Ges. Hydrobiol. Hydrogr., vol. 39, pp. 498–536.

KOCH, W.
1925 Die Geschichte der Binnenfischerei von Mitteleuropa. *In* Handbuch der Binnenfischerei Mitteleuropas 4a, Stuttgart.

KOESTLER, A.
1971 The Case of the Midwife Toad. New York, Random House. 187 pp.

KŘIŽENECKÝ, J. AND A. KŘIŽENECKA-PULÁNKOVÁ
1953 K otázce růstu, určování stáři a poměru pohlaví u okouna (*Perca fluviatilis* L.) [Growth, age determination and sex ratio in perch]. Mém. Soc. R. Sci. Bohème, vol. 15, pp. 1–26. [In Czech]

KUZNECOV, V. V.
1957 O tak nazyvajemom fenomene Li [On the so called phenomenon Lee]. Vop. Ikhtiol., vol. 8, pp. 143–154. [In Russian]

LEONHARDT, E.
1906 Der Karpfen. Geschichte, Naturgeschichte und wirtschaftliche Bedeutung unseres wichtigsten Zuchtfisches. Neudamm, J. Neumann. 104 pp.

LESZCZYŃSKA, W. AND L. BINIAKOWSKI
1967 Chów sazana w ośrodku zarybieniowym PZW Otorowo [Breeding of the wild carp in the sport fisheries stocking centre Otorowo]. Gospod. Rybna, vol. 19, no. 2, pp. 19–20. [In Polish]

LINDBERG, G. U.
1962 O sviazi kontinentov Evropy i Ameriki [About the unity of European and American continents]. Sov. rybochoz. issl. v sev.-zap. tchasti Atlan. ok., VNIRO (Moscow), pp. 69–82. [In Russian]

LÜHMANN, M. AND H. MANN
1957 Über Hungerformen beim Brassen. Der Fischwirt, vol. 8.

MAAR, A.
1960 The introduction of carp in Africa south of the Sahara. *In* Third Symposium on Hydrobiology and Inland Fisheries Problems of Major Lakes, Lusaka, CCTA, pp. 204–210.

ALBERTUS MAGNUS (Saint, Bishop of Ratisbon)
1861 De Animalibus libri XXVI. Beiträge zur Geschichte der Philosophie des Mittelalters (Münster), vol. 15 & 16, pp. 1–1664.

MARK, M.
1966 Carp breeding in drainage water. Bamidgeh, vol. 18, pp. 51–54.

MCCRIMMON, H. R.
1968 Carp in Canada. Bull. Fish. Res. Bd. Can., no. 165. pp. 1–93.

MIŠIK, V.
1958 Biometrika dunajského kapra (*Cyprinus carpio carpio* L.) z dunajského systému na Slovensku [Biometry of the Danube wild carp (*Cyprinus carpio carpio* L.) of the Danube basin in Slovakia]. Biol. práce, vol. 4, no. 6, pp. 55–125. [In Slovak, with Russian and German summaries].

MIŠIK, V. AND V. TUČA
1965 Posudzovanie exteriéru dunajského kapra so zretel'om na výber plemenného materiálu [Evaluation of the exterior of the Danubian carp with regard to the selection of the breeding material]. Pol'nohospodárstvo, vol. 11, pp. 35–44. [In Slovak, with Russian and German summaries].

MORCINEK, P.
1909 Geschichte des Dubischverfahrens. Land-und Forstwirtschaftliche Gesellschaft, Troppau.

MOVTCHAN, V. A.
1966 Zhinzn i neres ryb [The life and breeding of fishes]. Moscow, Kolos. [In Russian]

NAKAMURA, N.
1948 On the relation between salinity contents of the water and living condition and productivity of carps in ponds near the sea. Tokyo Univ. Physiograph. Soc. Res. Inst. Bull., vol. 1.

PATERCULUS, VELLEIUS, M.
1924 Compendium of Roman History; Res gestae divi Augusti, with an English translation by F. W. Shipley. London, W. Heinemann. 431 pp.

PELIKAN, O.
1960 Slovensko a Rímske Impérium [Slovakia and Roman Empire]. Bratislava, Slovenské vydavatel'stvo Krasnej literatúry. 348 pp. [In Slovak]

PERLBACH, M.
1881–1882 Pommerellisches Urkundenbuch. 2 vols. Danzig, Westpreussischer Geschichtsverein.

PLINIUS, S. C.
1635 Historiae naturalis. Lugduni Batavorum.

POPOVA, O. A.
1967 The "predator-prey" relationship among fish (A survey of Soviet papers). *In* Gerking, S. D., ed. The biological basis of freshwater fish production. Oxford, Blackwell Scientific, pp. 359–376.

RANDIK, A.
1967 Ochrana živočíchov v praxi. K vyhlaškam o ochrane vol'ne žijúcich živočíchov č. 80/1965 a č. 125/1965 [Rules for the protection of wildlife – Slovak National Council]. Ochrana Fauny, vol. 1, pp. 42–53. [In Slovak]

RUDZIŃSKI, E.
1961 Vergleichende Untersuchungen über den Wildkarpfen der Donau und den Teichkarpfen. Z. Fisch., vol. 10, pp. 105–135.
1962 Fragmenty historii europejskiego karpia hodowlanego [Fragments of the history of the European domestic carp]. Gospod. Rybna, vol. 14, pp. 3–5. [In Polish]

RUETIMEYER, L.
1860 Untersuchung der Thierreste aus den Pfahlbauten der Schweiz. Zurich, Mittl. Antiquarische Gesellschaft.

SEGERSTRÅLE, C.
1932 Über die jährlichen Zuwachszonen der Schuppen und Beziehungen zwischen Sommertemperatur und Zuwachs bei *Abramis brama* sowie einigen anderen Cypriniden in Süd-Finland 1911–1930. Acta Zool. Fenn., vol. 13, pp. 1–42.

STEFFENS, W.
1958 Der Karpfen. Wittenberg, A. Ziemsen. 90 pp.
1959 Wachstum und Körperform ungarischer Wildkarpfen in deutschen Teichen. Dt. Fisch Ztg, Radebeul, vol. 6, pp. 213–218.
1964 Vergleichende anatomisch-physiologische Untersuchungen an Wild – und Teichkarpfen (*Cyprinus carpio* L.). Ein Beitrag zur Beurteilung der Zuchtleistungen beim Deutschen Teichkarpfen. Z. Fisch., vol. 12, pp. 725–800.

STEFFENS, W.
1967 Das Domestikationsproblem beim Karpfen (*Cyprinus carpio* L.). Verh. int. Verein. theor. angew. Limnol., vol., 16, pp. 1441–1448.

STRABO, A.
1917–1933 The Geography of Strabo, with an English translation by H. L. Jones. Based in part on the unfinished version of J. R. S. Sterrett. 8 vols. London, W. Heinemann.

STROJNOWSKI, S.
1609 Opisanie porządku stawowego y przestróg niektórych domowego gospodarstwa ... *In* Z. Gawarecki and A. Kohn, 1860. Polskie stawowe gospodarstwo ... [Description of pond order & Polish pond management ...]. Warszawa. [In ancient Polish]

STRUMIEŃSKI, O.
1573 O sprawie, sypaniu, wymierzaniu i rybieniu stawów [On management, construction, measuring and stocking of ponds]. F. Kucharzewski, 1897, Cracow. 87 pp. [In ancient Polish]

TACITUS, P. C.
1836 The Works of Cornelius Tacitus, translated by A. Murphy. Philadelphia, T. Wardle. 742 pp.

TALBOT, L.M., W. J. A. PAYNE, H. P. LEDGER, L. D. VERDCOURT AND M. H. TALBOT
1965 The meat production potential of wild animals in Africa; a review of biological knowledge. Commonwealth Bureau of Animal Breeding and Genetics Technical Communication, no. 16, Farnham Royal, Bucks., Eng., Commonwealth Agricultural Bureau. 42 pp.

TAMURA, T.
1961 Carp cultivation in Japan. *In* Borgström, G., ed. Fish as Food, vol. 1. Production, biochemistry and microbiology. New York, Academic Press, pp. 103–120.

TAYLOR, C. R.
1970 Dehydration and heat: effects on temperature regulations of East African ungulates. Am. J. Physiol., vol. 219, no. 4, pp. 1136–1139.
1972 Ranching arid lands: Physiology of wild and domestic ungulates in the desert. Botswana Notes and Records, sp. ed. no. 1, Proc. Conf. on Sustained Production from Semi-Arid Areas, Oct. 1971, Gaberone, pp. 167–192. [Mimeographed]

TCHEN, S. C.
1956 A history of the domestication and the factors of the varietal formation of the common goldfish, *Carassius auratus.* Scientia Sin., vol. 2, pp. 287–322.

THIENEMANN, A.
1950 Verbreitungsgeschichte der Süsswassertierwelt Europas. Die Binnengewässer Band 18. Stuttgart, E. Schweizerbartsche. 809 pp.

TUČA, V.
1958 Štúdium genetickej (exteriérovej) čistoty dunajského kapra [Study of genetical (exterior) purity of the Danube carp]. Lab. Fish. Res. Final Rept., Bratislava. Typescript. [In Slovak]

VALCANOVER, F.
1960 All the paintings of Titian. Hawthorn Books, Inc., New York. Part 2, Plate 127–128.

VARRO, M. T.
1912 Varro on Farming. M. Terenti Varronis Rerum Rusticarum libri tres, translated by L. Storr-Best. London, G. Bell. 375 pp.

VOLF, J.
1965 Kralik. Svetem zvířat. Sv. 3. Domaci zvířata [The rabbit. *In* Hanzak, J., J. Volf and L. J. Dobroruka. Through Animal World, Vol. 3. Domestic Animals.] Prague, SNDK. [In Czech]

VOOREN, C. M.
1972 Ecological aspects of the introduction of fish species into natural habitats in Europe, with special reference to the Netherlands. J. Fish Biol., vol. 4, no. 4, pp. 565–583.

WUNDER, W.
1949 Fortschrittliche Karpfenteichwirtschaft. Stuttgart.

ZAUNICK, R.
1925 Tritt der Karpfen schon im Diluvium Norddeutschlands auf ? Mitteilg. d. Fischereivereine Brandenburg-Pommern etc., vol. 17, pp. 80–83.

Резюме

Для установления исторического начала разведения карпа были использованы исторические, зоогеографические, морфологические и физиологические сведения. Предки дикого карпа (сазана) вероятно берут свое начало в Средней Азии и потом естественным путём распространяются на восток, в Китай и на запад, до реки Дунай. На основании исторических данных можно предполагать, что римляне первые начали разведение сазана из реки Дунай. И эта традиция выдерживания сазана в *piscinae* продолжалась в монастырях в Средние Века. Выносливость дикого карпа из реки Дунай продемонстрирована на опыте голодования этой рыбы. Разведение карпа в Китае вначале было независимым от подобной деятельности Европы. Убеждающие доказательства указывают на то, что дикий карп реки Дунай является предком многих современных культивированных видов карпа. Это является важным фактором в пользу сохранения этой рыбы для дальнейшего употребления.

Поскольку доледниковые остатки карпа в сфайных постройках не признают из-за неточности их даты, поэтому предполагают, что дикий карп происходит из Средней Азии, откуда он распространился естественным путём на восток, в Китай и смежные края и на запад, в последнюю послеледниковую эпоху, до самого Дуная. В доказательство того, что это не произошло раньше приводится факт, что в Плейстоценовую эпоху после отхода континентальных ледников *C. carpio* не появился вместе с другими видами рыб, которые проникли в Скандинавию, на Британские Острова и в Северную Америку. Если бы в этот период карп появился в восточной Азии, то он возможно появился бы и в Северной Америке, естественным путём, вместе со щукой и другими пресноводными рыбами, которые проникли туда через Берингов пролив.

В доказательство естественного появления дикого карпа в Дунае взяты следующие факты: разными именами называли карпа местные жители; одним и тем же кельтийским именем называли эту рыбу вне её естественных западных границ пребывания; сведения из Средневековых летописей о распространении карпа; из современных исследований о диком карпе, живущем в реке Дунай. Дифференциация дикого карпа в далеком прошлом в Средней Азии и появление неизменившихся поодиноких групп этой рыбы в восточных и западных районах наводит на мысль о том, что местом происхождения карпа должна быть Средняя Азия.

Отдельная глава посвящается опыту, исследующему способность дикого карпа реки Дунай выжить в условиях голода. Этот опыт подтверждает выносливость этой рыбы и объясняет почему карпа разводили люди для пищи.

Наконец, из записей Римлян и из выкопанных предметов искусства, изображающих карпа, напрашивается гипотеза о том, что впервые эта рыба попала в рыбные садки Римлян из тех мест, где проходила Янтарная Дорога — самая главная дорога тех времён, пересекающая реку Дунай. Это были места регулярно наводнённых лугов в устье реки Моравы, около древнего Карнунтума и Замка Девина, где обыкновенно стаи карпа метали икру. Патриции, командиры римских легионов, стоявших в этих местах, завезли карпа в Рим, как эпикурийский деликатес потому, что это была одна из самых выносливых видов рыб, которая смогла пережить примитивную технику транспорта и разведение рыбы тех времён. После упадка и разрушения Римской Империи техника разведения карпа в римских рыбных садках не была заброшена. Это дело продолжали християнские миссионеры и монахи, нуждающиеся в запасах рыбы во время религиозных постов. Миссионеры и монахи знали из латинских записей, а также из опыта южных стран, откуда они вышли о том, что карп подходит для разведения и они занимались этим делом в местах своего нового поселения. С концом Средних Веков разведение карпа было уже рутиной не только в монастырях, но и в частных имениях.

История разведения карпа в Европе, начинающаяся диким карпом из реки Дунай (*Cyprinus carpio carpio*), делится на пять периодов: 1 — период в котором первыми начали разводить карпа Римляне, на запад от реки Дунай, в первом и до четвертого столетия нашей эры; 2 — период спорадического начала разведения карпа в пятом и шестом столетиях; 3 — период массового разведения карпа и стремления к разведению этой рыбы в большом количестве в специально заведенных прудах в седьмом и по тринадцатое столетия; 4 — период широко распространенного разведения карпа и случайное селекционирование в четырнадцатом и по шестнадцатое столетия; 5 — период усиленного разведения, целеустремленная селекция и начало разведения карпа в Северной Америке, Австралии, Дальнем Востоке и в Африке с семнадцатого столетия. Возможно в Китае разведение карпа началось из местного подвида *C. c. haematopterus* 500 лет раньше, но не зависимо от Европы.

Здесь подчеркивается самое важное свойство, по которому дикий карп реки Дунай отличается от теперешнего культивированного карпа и дается ссылка на разительное сходство между разведением карпа и приручиванием кролика. Также обращается внимание на потребность сохранения последних диких стай карпа в предгорьях реки Дунай и предохранение дикого предка культивированного карпа в его естественной среде. И самое главное, повторение судьбы вымершего тура, предка крупного рогатого скота, не должно быть допустимо. Как крупный рогатый скот, так и культивированный карп в диком состоянии является рыбой плохого качества. Это, возможно, и является причиной того, что карп стал не популярным в Америке после его интродукции.